Polymer Chemistry

Polymer Chemistry

Edited by
Walter O'Hara

Larsen & Keller
www.larsen-keller.com

Polymer Chemistry
Edited by Walter O'Hara
ISBN: 978-1-63549-228-6 (Hardback)

☐ Larsen & Keller

Published by Larsen and Keller Education,
5 Penn Plaza,
19th Floor,
New York, NY 10001, USA

Cataloging-in-Publication Data

Polymer chemistry / edited by Walter O'Hara.
   p. cm.
Includes bibliographical references and index.
ISBN 978-1-63549-228-6
1. Polymers. 2. Polymerization. 3. Polymer engineering.
4. Chemical engineering. I. O'Hara, Walter.
QD381 .P65 2017
547.7--dc23

The publisher's policy is to use permanent paper from mills that operate a sustainable forestry policy. Furthermore, the publisher ensures that the text paper and cover boards used have met acceptable environmental accreditation standards.

Printed and bound in the United States of America.

For more information regarding Larsen and Keller Education and its products, please visit the publisher's website www.larsen-keller.com

Table of Contents

Preface

This book elucidates the concepts and innovative models around prospective developments with respect to polymer chemistry. It describes in detail the various concepts and theories related to the field. Polymer chemistry refers to that branch of chemistry which deals with the study of the different characteristics, functions and structures of synthetic polymers. This field encompasses the concepts of other related areas like polymer physics, polymer science, and polymer engineering. The topics included in this book are of utmost significance and bound to provide incredible insights to readers. It attempts to assist those with a goal of delving into the field of polymer chemistry. For someone with an interest and eye for detail, this book covers the most significant developments related to the field of polymer chemistry.

Given below is the chapter wise description of the book:

Chapter 1- Polymers are large molecules which are composed of subunits. Polymer chemistry deals with the structure, design and manufacture of these polymers. This chapter is an overview of the subject matter incorporating all the major aspects of polymer chemistry.

Chapter 2- The distribution of monomers in a copolymer is known as the Mayo-Lewis equation. Apart from the Mayo-Lewis equation, the other important theories of polymer chemistry are Flory-Stockmayer theory, Flory-Huggins solution theory, Hoffman nucleation theory, polymer field theory etc. This text elucidates the crucial theories of polymer chemistry.

Chapter 3- Large molecules that contain repeated subunits are polymers. A molecule that binds chemically to other molecules to form a polymer is known as a monomer. Polymers have a number of classes such as biopolymer, conductive polymer, copolymer, silicone and smart polymer. This chapter is an overview of the subject matter incorporating all the major aspects of polymer chemistry.

Chapter 4- Different polymers have different properties. Some of the most popular and common polymers manufactured are thermoplastic, polyethylene, polystyrene etc. They are manufactured keeping in mind various aspects such as use, strength and degradation. Polymers are best understood in confluence with the major topics listed in the following chapter.

Chapter 5- Polymerization has a number of forms. Some of these are step-growth polymerization, chain-growth polymerization and reversible-deactivation polymerization. The forms of chain growth polymerization such as, anionic addition polymerization, cationic polymerization and living polymerization are also explained in the following chapter.

Chapter 6- Polymer degradation is the change in the properties of a polymer. Some of these properties are the shape, tensile strength and shape of a polymer. These variations are usually unwanted. Some other aspects of polymer degradation listed in this chapter are UV degradation, thermal degradation of polymers, thermal depolymerization and photo-oxidation of polymers.

Indeed, my job was extremely crucial and challenging as I had to ensure that every chapter is informative and structured in a student-friendly manner. I am thankful for the support provided by my family and colleagues during the completion of this book.

Editor

Introduction to Polymer Chemistry

Polymers are large molecules which are composed of subunits. Polymer chemistry deals with the structure, design and manufacture of these polymers. This chapter is an overview of the subject matter incorporating all the major aspects of polymer chemistry.

Polymer chemistry (also called macromolecular chemistry) is the science of chemical synthesis and chemical properties of polymers or macromolecules. According to IUPAC recommendations macromolecules refer to the individual molecular chains and are the domain of chemistry. Polymers describe the bulk properties of polymer materials and belong to the field of polymer physics (a part of physics).

The different kinds of macromolecules include:
- Biopolymers produced by living organisms:
  - structural proteins: collagen, keratin, elastin and others
  - chemically functional proteins: enzymes, hormones, transport proteins and others
  - structural polysaccharides: cellulose, chitin and others
  - storage polysaccharides: starch, glycogen and others
  - nucleic acids: DNA, RNA
- Synthetic polymers used for plastics—fibers, paints, building materials, furniture, mechanical parts, adhesives:
  - thermoplastics: polyethylene, Teflon, polystyrene, polypropylene, polyester, polyurethane, polymethyl methacrylate, polyvinyl chloride, nylon, rayon, celluloid, silicone and others
  - thermosetting plastics: vulcanized rubber, Bakelite, Kevlar, epoxy and others.

Polymers are formed by polymerization of monomers. Chemists describe a polymer by its degree of polymerization, molar mass distribution, tacticity, copolymer distribution, the degree of branching, by its end-groups, crosslinks, and crystallinity. Chemists also study a polymer's thermal properties such as its glass transition temperature and melting temperature. Polymers in solution have special characteristics for solubility, viscosity and gelation.

History

Polymer chemistry started by studying the long fibers in plants. The work of Henri Braconnot in 1777 and the work of Christian Schönbein in 1846 led to the discovery of nitrocellulose. Nitrocellulose treated with camphor makes celluloid. Chemists dissolve celluloid in ether or acetone to make collodion. Doctors have used collodion as a wound dressing since the U.S. Civil War. Cellulose acetate was first prepared in 1865. In 1834, Friedrich Ludersdorf and Nathaniel Hayward independently discovered that adding sulfur to raw natural rubber (polyisoprene) helped prevent the material from becoming sticky. In 1844 Charles Goodyear received a U.S. patent for vulcanizing rubber with sulfur and heat. Thomas Hancock had received a patent for the same process in the UK the year before.

In 1884, Hilaire de Chardonnet started the first artificial fiber factory based on regenerated cellulose, or viscose rayon, as a substitute for silk, but it was very flammable. In 1907 Leo Baekeland invented the first synthetic polymer, a thermosetting phenol-formaldehyde resin called Bakelite. Around the same time, Hermann Leuchs reported the synthesis of N-carboxyanhydrides and their high molecular weight products upon reaction with nucleophiles. But Leuchs did not call them polymers, possibly due to the strong views held by Emil Fischer, his direct supervisor, denying the possibility of any covalent molecule exceeding 6,000 daltons. Cellophane was invented in 1908 by Jocques Brandenberger who squirted sheets of viscose rayon into an acid bath.

In 1922, Hermann Staudinger (a German chemist) proposed that polymers were long chains of atoms held together by covalent bonds. He also proposed to name these compounds "macromolecules". Before that, scientists believed that polymers were clusters of small molecules (called colloids), without definite molecular weights, held together by an unknown force. Staudinger received the Nobel Prize in Chemistry in 1953.

Wallace Carothers invented the first synthetic rubber called neoprene in 1931. Neoprene was the first polyester. Carothers went on to invent nylon, a true silk replacement, in 1935. Paul Flory was awarded the Nobel Prize in Chemistry in 1974 for his work on polymer random coil configurations in solution in the 1950s. Stephanie Kwolek developed an aramid, or aromatic nylon named Kevlar, patented in 1966.

There are now a large number of commercial polymers. They include composite materials such as carbon fiber-epoxy, polystyrene-polybutadiene (HIPS), acrylonitrile-butadiene-styrene (ABS). Chemists design commercial polymers to combine the best properties of their various components. For example, special polymers used in automobile engines are designed to work at high temperatures.

It took a long time before universities introduced teaching and research programs in polymer chemistry. An "Institut fur Makromolekulare Chemie was founded in 1940 in Freiburg, Germany under the direction of Hermann Staudinger. In America a "Polymer Research Institute" (PRI) was established in 1941 by Herman Mark at the Polytechnic Institute of Brooklyn (now Polytechnic Institute of NYU). Several hundred graduates of PRI played an important role in the US polymer industry and academia. Other PRI's were founded in 1961 by Richard S. Stein at the University of Massachusetts, Amherst, in 1967 by Eric Baer at Case Western Reserve University and in 1988 at the University of Akron.

References

- "Macromolecule". IUPAC. http://old.iupac.org/reports/1996/6812jenkins/molecules.html#1.1. Retrieved 2011-09-05.

- "The Early Years of Artificial Fibres". The Plastics Historical Society. http://www.plastiquarian.com/index.php?articleid=286. Retrieved 2011-09-05.

- "History of Cellophane". about.com. http://inventors.about.com/od/cstartinventions/a/Cellophane.htm. Retrieved 2011-09-05.

- "The History of Kevlar". about.com. http://inventors.about.com/library/inventors/blkevlar.htm. Retrieved 2011-09-05.

Theories of Polymer Chemistry

The distribution of monomers in a copolymer is known as the Mayo-Lewis equation. Apart from the Mayo-Lewis equation, the other important theories of polymer chemistry are Flory-Stockmayer theory, Flory-Huggins solution theory, Hoffman nucleation theory, polymer field theory etc. This text elucidates the crucial theories of polymer chemistry.

Flory-Stockmayer Theory

Flory-Stockmayer Theory is a theory governing the cross-linking and gelation of step-growth polymers:.. The Flory-Stockmayer theory represents an advancement from the Carothers equation, allowing for the identification of the gel point for polymer synthesis not at stoichiometric balance. The theory was initially conceptualized by Paul Flory in 1941 and then was further developed by Walter Stockmayer in 1944 to include cross-linking with an arbitrary initial size distribution.

History

Gelation occurs when a polymer forms large interconnected polymer molecules through cross-linking. In other words, polymer chains are cross-linked with other polymer chains to form an infinitely large molecule, interspersed with smaller complex molecules, shifting the polymer from a liquid to a network solid or gel phase. The Carothers equation is an effective method for calculating the degree of polymerization for stoichiometrically balanced reactions. However, the Carothers equation is limited to branched systems, describing the degree of polymerization only at the onset of cross-linking. The Flory-Stockmayer Theory allows for the prediction of when gelation occurs using percent conversion of initial monomer and is not confined to cases of stoichiometric balance. Additionally, the Flory-Stockmayer Theory can be used to predict whether gelation is possible through analyzing the limiting reagent of the step-growth polymerization.

Flory's Assumptions

In creating the Flory-Stockmayer Theory, Flory made three assumptions that affect the accuracy of this model. These assumptions were

1. All functional groups on a branch unit are equally reactive

2. All reactions occur between A and B

3. There are no intramolecular reactions

As a result of these assumptions, a conversion slightly higher than that predicted by the Flory-Stockmayer Theory is commonly needed to actually create a polymer gel. Since steric hindrance effects prevent each functional group from being equally reactive and intramolecular reactions do occur, the gel forms at slightly higher conversion

General Case

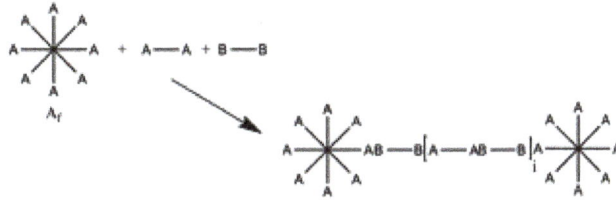

A general image of a multifunctional branch unit, A_f, reacting with bifunctional monomers with A and B functional groups to form a step-growth polymer.

The Flory-Stockmayer Theory predicts the gel point for the system consisting of three types of monomer units

   linear units with two A-groups (concentration c_1),

   linear units with two B groups (concentration c_2),

   branched A units (concentration c_3).

The following definitions are used to formally define the system

   f is the number of reactive functional groups on the branch unit (i.e. the functionality of that branch unit)

   p_A is the probability that A has reacted (conversion of A groups)

   p_B is the probability that B has reacted (conversion of B groups)

$$\rho = \frac{fc_3}{2c_1 + fc_3}$$ is the ratio of number of A groups in the branch unit to the total number of A groups

$$r = \frac{2c_1 + fc_3}{2c_2} = \frac{p_B}{p_A}$$ is the ratio between total number of A and B groups. So that $p_B = rp_A$.

The theory states that the gelation occurs only if $\alpha > \alpha_c$, where

$$\alpha_c = \frac{1}{f-1}$$

is the critical value for cross-linking and α is presented as a function of p_A,

$$\alpha(p_A) = \frac{rp_A^2\rho}{1 - rp_A^2(1-\rho)}$$

or, alternatively, as a function of p_B,

$$\alpha(p_B) = \frac{p_B^2 \rho}{r - p_B^2(1-\rho)}.$$

One may now substitute expressions for r, ρ into definition of α and obtain the critical values of $p_A, (p_B)$ that admit gelation. Thus gelation occurs if

$$p_A > \sqrt{\frac{\alpha_c}{r(\alpha_c + \rho - \alpha_c \rho)}}.$$

alternatively, the same condition for p_B reads,

$$p_B > \sqrt{\frac{r\alpha_c}{\alpha_c + \rho - \alpha_c \rho}}$$

The both inequalities are equivalent and one may use the one that is more convenient. For instance, depending on which conversion p_A or p_B is resolved analytically.

Trifunctional A Monomer with Difunctional B Monomer

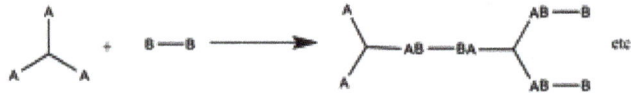

A trifunctional branch unit with A functional group reacting with a bifunctional branch unit with a B functional group, forming a continuous step-growth polymer molecule.

$$\alpha_c = \frac{1}{f-1} = \frac{1}{3-1} = \frac{1}{2}$$

Since all the A functional groups are from the trifunctional monomer, $\rho = 1$ and

$$\alpha = \frac{\dfrac{p_B^2 \rho}{r}}{1 - \dfrac{p_B^2}{r(1-\rho)}} = \frac{p_B^2}{r}$$

Therefore, gelation occurs when

$$\frac{p_B^2}{r} > \alpha_c$$

or when,

$$p_B > \sqrt{\frac{r}{2}}$$

Similarly, gelation occurs when

$$p_A > \sqrt{\frac{1}{2r}}$$

Flory–Huggins Solution Theory

Flory–Huggins solution theory is a mathematical model of the thermodynamics of polymer solutions which takes account of the great dissimilarity in molecular sizes in adapting the usual expression for the entropy of mixing. The result is an equation for the Gibbs free energy change ΔG_m for mixing a polymer with a solvent. Although it makes simplifying assumptions, it generates useful results for interpreting experiments.

The thermodynamic equation for the Gibbs energy change accompanying mixing at constant temperature and (external) pressure is

$$\Delta G_m = \Delta H_m - T\Delta S_m$$

A change, denoted by Δ, is the value of a variable for a solution or mixture minus the values for the pure components considered separately. The objective is to find explicit formulas for ΔH_m and ΔS_m, the enthalpy and entropy increments associated with the mixing process.

The result obtained by Flory and Huggins is

$$\Delta G_m = RT[n_1 \ln \phi_1 + n_2 \ln \phi_2 + n_1 \phi_2 \chi_{12}]$$

The right-hand side is a function of the number of moles n_1 and volume fraction ϕ_1 of solvent (component 1), the number of moles n_2 and volume fraction ϕ_2 of polymer (component 2), with the introduction of a parameter χ to take account of the energy of interdispersing polymer and solvent molecules. R is the gas constant and T is the absolute temperature. The volume fraction is analogous to the mole fraction, but is weighted to take account of the relative sizes of the molecules. For a small solute, the mole fractions would appear instead, and this modification is the innovation due to Flory and Huggins. In the most general case the mixing parameter, χ, is a free energy parameter, thus including an entropic component.

Derivation

We first calculate the *entropy* of mixing, the increase in the uncertainty about the locations of the molecules when they are interspersed. In the pure condensed phases — solvent and polymer — everywhere we look we find a molecule. Of course, any notion of "finding" a molecule in a given location is a thought experiment since we can't actually examine spatial locations the size of molecules. The expression for the entropy of mixing of small molecules in terms of mole fractions is no longer reasonable when the solute is a macromolecular chain. We take account of this dissymmetry in molecular sizes by assuming that individual polymer segments and individual solvent molecules occupy sites on a lattice. Each site is occupied by exactly one molecule of the solvent or by one monomer of the polymer chain, so the total number of sites is

$$N = N_1 + xN_2$$

N_1 is the number of solvent molecules and N_2 is the number of polymer molecules, each of which has x segments.

From statistical mechanics we can calculate the entropy change, the increase in spatial uncertainty, as a result of mixing solute and solvent.

$$\Delta S_m = -k[N_1 \ln(N_1/N) + N_2 \ln(xN_2/N)]$$

where k is Boltzmann's constant. Define the lattice *volume fractions* ϕ_1 and ϕ_2

$$\phi_1 = N_1/N$$

$$\phi_2 = xN_2/N$$

These are also the probabilities that a given lattice site, chosen at random, is occupied by a solvent molecule or a polymer segment, respectively. Thus

$$\Delta S_m = -k[N_1 \ln \phi_1 + N_2 \ln \phi_2]$$

For a small solute whose molecules occupy just one lattice site, x equals one, the volume fractions reduce to molecular or mole fractions, and we recover the usual equation from ideal mixing theory.

In addition to the entropic effect, we can expect an *enthalpy* change. There are three molecular interactions to consider: solvent-solvent w_{11}, monomer-monomer w_{22} (not the covalent bonding, but between different chain sections), and monomer-solvent w_{12}. Each of the last occurs at the expense of the average of the other two, so the energy increment per monomer-solvent contact is

$$\Delta w = w_{12} - \frac{1}{2}(w_{22} + w_{11})$$

The total number of such contacts is

$$xN_2 z\phi_1 = N_1\phi_2 z$$

where z is the coordination number, the number of nearest neighbors for a lattice site, each one occupied either by one chain segment or a solvent molecule. That is, xN_2 is the total number of polymer segments (monomers) in the solution, so xN_2z is the number of nearest-neighbor sites to *all* the polymer segments. Multiplying by the probability ϕ_1 that any such site is occupied by a solvent molecule, we obtain the total number of polymer-solvent molecular interactions. An approximation following mean field theory is made by following this procedure, thereby reducing the complex problem of many interactions to a simpler problem of one interaction.

The enthalpy change is equal to the energy change per polymer monomer-solvent interaction multiplied by the number of such interactions

$$\Delta H_m = N_1\phi_2 z\Delta w$$

The polymer-solvent interaction parameter *chi* is defined as

$$\chi_{12} = z\Delta w / kT$$

It depends on the nature of both the solvent and the solute, and is the only *material-specific* parameter in the model. The enthalpy change becomes

$$\Delta H_m = kTN_1\phi_2\chi_{12}$$

Assembling terms, the total free energy change is

$$\Delta G_m = RT[n_1 \ln \phi_1 + n_2 \ln \phi_2 + n_1 \phi_2 \chi_{12}]$$

where we have converted the expression from molecules N_1 and N_2 to moles n_1 and n_2 by transferring Avogadro's number N_A to the gas constant $R = kN_A$.

The value of the interaction parameter can be estimated from the Hildebrand solubility parameters δ_a and δ_b

$$\chi_{12} = V_{seg}(\delta_a - \delta_b)^2 / RT$$

where V_{seg} is the actual volume of a polymer segment.

This treatment does not attempt to calculate the conformational entropy of folding for polymer chains. The conformations of even amorphous polymers will change when they go into solution, and most thermoplastic polymers also have lamellar crystalline regions which do not persist in solution as the chains separate. These events are accompanied by additional entropy and energy changes.

It should be noted that in the most general case the interaction Δw and the ensuing mixing parameter, χ, is a free energy parameter, thus including an entropic component. This means that aside to the regular mixing entropy there is another entropic contribution from the interaction between solvent and monomer. This contribution is sometimes very important in order to make quantitative predictions of thermodynamic properties.

More advanced solution theories exist, such as the Flory-Krigbaum theory.

Oligomer-gel Solutions Theory

Recently a team of scientists from the University of Durham (UK) introduced a new form of the Flory-Huggins free energy for a solution where there is an oligomer with a cross-linked polymer like a reticulated gel. They considered the mixing entropy for the oligomer and the elastic entropy for the reticulated gel keeping the mean field interaction through χ. They found that increasing the elastic modulus of the cross-linked polymer it is possible to increase the critical value of the mean field over which there is phase separation. This theory introduces a new way of controlling the phase separation in such a system. Furthermore they used this theory for studying the wetting and the migration of the oligomer specie to a surface.

Hoffman Nucleation Theory

Hoffman nucleation theory is a theory developed by John D. Hoffman and coworkers in the 1970s and 80s that attempts to describe the crystallization of a polymer in terms of the kinetics and thermodynamics of polymer surface nucleation. The theory introduces a model where a surface of completely crystalline polymer is created and introduces surface energy parameters to describe the process. Hoffman nucleation theory is more of a starting point for polymer crystallization theory and is better known for its fundamental roles in the Hoffman-Weeks Lamellar Thickening and Lauritzen-Hoffman Growth Theory.

Polymer Morphology

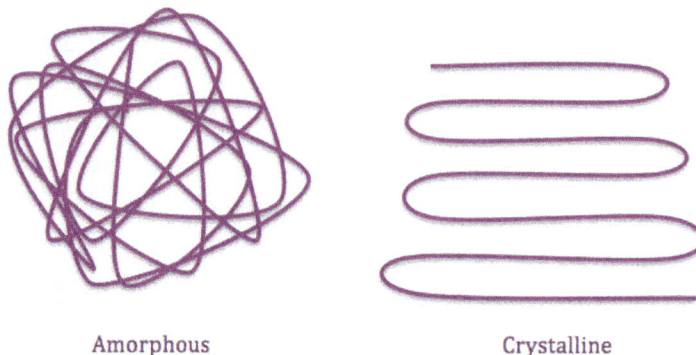

Amorphous   Crystalline

Amorphous regions lack the energy needed to order into folded regions such as those seen in the Crystalline state

Polymers contain different morphologies on the molecular level which give rise to their macro properties. Long range disorder in the polymer chain is representative of amorphous solids, and the chain segments are considered amorphous. Long range polymer order is similar to crystalline material, and chain segments are considered crystalline.

The thermal characteristics of polymers are fundamentally different from those of most solid materials. Solid materials typically have one melting point, the T_m, above which the material loses internal molecular ordering and becomes a liquid. Polymers have both a melting temperature T_m and a glass transition temperature T_g. Above the T_m, the polymer chains lose their molecular ordering and exhibit reptation, or mobility. Below the T_m, but still above the T_g, the polymer chains lose some of their long-range mobility and can form either crystalline or amorphous regions. In this temperature range, as the temperature decreases, amorphous regions can transition into crystalline regions, causing the bulk material to become more crystalline over all. Below the T_g, molecular motion is stopped and the polymer chains are essentially frozen in place. In this temperature range, amorphous regions can no longer transition into crystalline regions, and the polymer as a whole has reached its maximum crystallinity.

	Temperature	Polymer Macroproperties	Polymer Morphology
T_m	$>T_m$	Liquid	No long-range order, polymer chains are fully amorphous and can reptate.
	$T_m > T > T_g$	Rubber	Polymer chains contain amorphous and crystalline regions and lower overall mobility.
T_g	$<T_g$	Glassy/Brittle	Polymer chains are stuck in place and no more crystallinity can be achieved.

Hoffman nucleation theory addresses the amorphous to crystalline polymer transition, and this transition can only occur in the temperature range between the T_m and T_g. The transition from an amorphous to a crystalline single polymer chain is related to the random thermal energy required to align and fold sections of the chain to form ordered regions titled lamellae, which are a subset of even bigger structures called spherulites. The Crystallization of polymers can be brought about by several different methods, and is a complex topic in itself.

Nucleation

Nucleation is the formation and growth of a new phase with or without the presence of external surface. The presence of this surface results in heterogeneous nucleation whereas in its absence homogeneous nucleation occurs. Heterogeneous nucleation occurs in cases where there are pre-existing nuclei present, such as tiny dust particles suspended in a liquid or gas or reacting with a glass surface containing SiO_2. For the process of Hoffman nucleation and its progression to Lauritzen-Hoffman Growth Theory, homogeneous nucleation is the main focus. Homogeneous nucleation occurs where no such contaminants are present and is less commonly seen. Homogeneous nucleation begins with small clusters of molecules forming from one phase to the next. As the clusters grow, they aggregate through the condensation of other molecules. The size continues to increase and ultimately form macroscopic droplets (or bubbles depending on the system).

Nucleation is often described mathematically through the change in Gibbs free energy of n moles of vapor at vapor pressure P that condenses into a drop. Also the nucleation barrier, in polymer crystallization, consists of both enthalpic and entropic components that must be over come. This barrier consists of selection processes taking place in different length and time scales which relates to the multiple regimes later on. This barrier is the free energy required to overcome in order to form nuclei. It is the formation of the nuclei from the bulk to a surface that is the interfacial free energy. The interfacial free energy is always a positive term and acts to destabilize the nucleus allowing the continuation of the growing polymer chain. The nucleation continues as a favorable reaction.

Thermodynamics of Polymer Crystallization

The Lauritzen–Hoffman plot (right) models the three different regimes when $(\log G) + U^*/k(T-T_o)$ is plotted against $(T\Delta T)^{-1}$. It can be used to describe the rate at which secondary nucleation competes with lateral addition at the growth front among the different temperatures. This theory can be used to help understand the preferences of nucleation and growth based on the polymer's properties including its standard melting temperature.

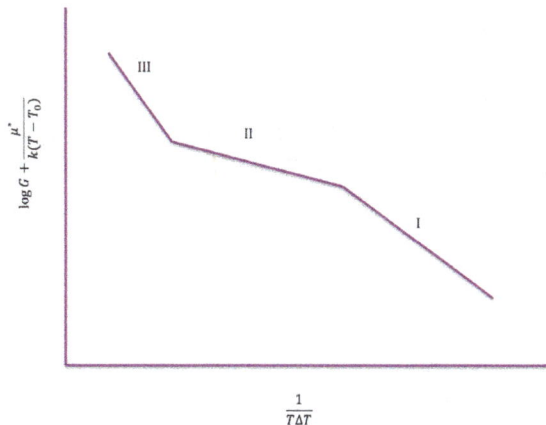

Lauritzen–Hoffman Plot detailing the Three Regimes of Secondary Nucleation

Lamellar Thickening (Hoffman–Weeks plot)

For many polymers, the change between the initial lamellar thickness at T_c is roughly the same as at T_m and can thus be modeled by the Gibbs–Thomson equation fairly well. However, since it implies that the lamellar thickness over the given supercooling range (T_m-T_c) is unchanged, and many homogeneous nucleation of polymers implies a change of thickness at the growth front, Hoffman and Weeks pursued a more accurate representation. In this regard, the Hoffman-Weeks plot was created and can be modeled through the equation

$$T_m = \frac{T_c}{\beta} + (1 - \frac{1}{\beta})T_m^\circ$$

where β is representative of a thickening factor given by $L = L_o \beta$ and T_c and T_m are the crystallization and melting temperatures, respectively.

Applying this experimentally for a constant β allows for the determination of the equilibrium melting temperature, T_m° at the intersection of T_c and T_m.

Kinetics of Polymer Crystallization

The crystallization process of polymers does not always obey simple chemical rate equations. Polymers can crystallize through a variety of different regimes and unlike simple molecules, the polymer crystal lamellae have two very different surfaces. The two most prominent theories in polymer crystallization kinetics are the Avrami equation and Lauritzen-Hoffman Growth Theory.

Lauritzen-Hoffman Growth Theory

The Lauritzen-Hoffman growth theory breaks the kinetics of polymer crystallization into ultimately two rates. The model breaks down into the addition of monomers onto a growing surface. This initial step is generally associated with the nucleation of the polymer. From there, the kinetics become the rate which the polymer grows on the surface, or the lateral growth rate, in comparison with the growth rate onto the polymer extending the chain, the secondary nucleation rate. These two rates can result in three situations.

Three Regimes of Crystallization Kinetics

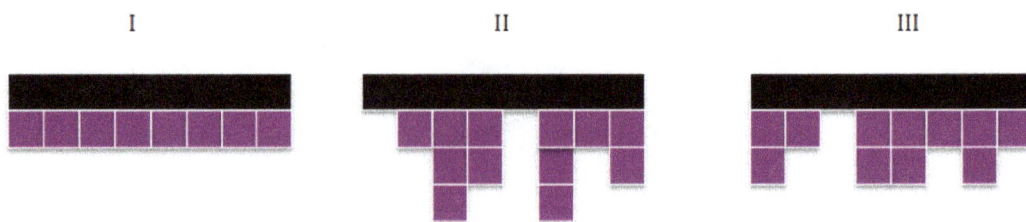

For Regime I, the growth rate on the front laterally, referred to as g, is the rate-determining step (RDS) and exceeds the secondary nucleation rate, i. In this instance of $g \gg i$, monolayers are formed one at a time so that if the substrate has a length of L_p and thickness, b, the overall linear growth can be described through the equation

$$G_I = biL_p$$

and the rate of nucleation in specific can further be described by

$$G_{I,n} = e^{-(K_g/T\Delta T)}$$

with K_g equal to

$$K_g = \frac{4b\sigma_l\sigma_f T_m^0}{k\Delta h}$$

where

- σ_l is the lateral/lamellae surface free energy per unit area

- σ_f is the fold surface free energy per unit area

- T_m^0 is the equilibrium melting temperature

- k is equal to Boltzmann Constant

- Δh is equal to the change of enthalpy of fusion per repeat unit at the standard temperature (Also known as the Latent Heat of Fusion per repeat unit)

This shows that in Region I, lateral nucleation along the front successfully dominates at temperatures close to the melting temperature, however at more extreme temperatures other forces such as diffusion can impact nucleation rates.

In Regime II, the lateral growth rate is either comparable or smaller than the nucleation rate $g \leq i$, which causes secondary (or more) layers to form before the initial layer has been covered. This allows the linear growth rate to be modeled by

$$G_{II} = b\sqrt{ig}$$

Using the assumption that g and i are independent of time, the rate at which new layers are formed can be approximated and the rate of nucleation in regime II can be expressed as

$$G_{II,n} = e^{-(K_g'/T\Delta T)}$$

with K_g' equal to about 1/2 of the K_g from Regime I,

$$K_g' = \frac{2b\sigma_l\sigma_f T_m^0}{k\Delta h}$$

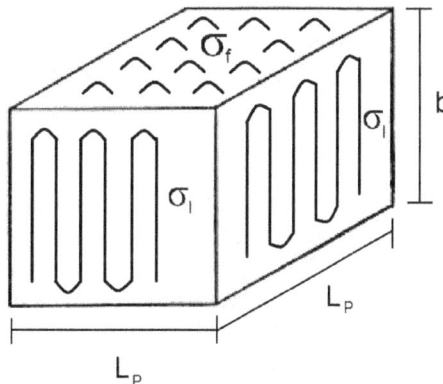

Diagram of a crystalline polymer lamellae

Lastly, Regime III in the L-H model depicts the scenario where lateral growth is inconsequential to the overall rate, since the nucleation of multiple sites causes $i >> g$. This means that the growth rate can be modeled by the same equation as Regime I,

$$G_{III} = biL_p = G_{III}^{\circ} e^{-U^*/k(T-T_0)-(K_g/T\Delta T)}$$

where G_{III}° is the prefactor for Regime III and can be experimentally determined through applying the Lauritzen–Hoffman Plot.

Polyethylene Crystallization Kinetics

A reza's crystallization depends on the time it takes for layers of its chains to fold and orient themselves in the same direction. This time increases with a molecule's weight and branching. The table below shows that the growth rate is higher for Sclair 14B.1 than Sclair 2907 (20%), where 2907 is less highly branched than 14B.1. Here Gc is the crystal growth rate, or how quickly it orders itself depending on the layers, and t is the time it takes to order.

Polymer	Growth Temp (ᵒC)	G_c (μm*min^{-1})	t (ms)
Sclair 2907 (20%)	119	3.5-6.8	4.4-8.6
Sclair 14B.1	119	~0.2	~150

Further Testing and Applications

Many additional tests have since been run to apply and compare Hoffman's principles to reality. Among the experiments done, some of the more notable secondary nucleation tests are briefly explained in the table below.

Secondary nucleation testing	Experimental results observed
Potassium chloride (KCl)	Secondary nuclei form at a rate proportional to the degree of supercooling (above certain levels of agitation) and achieve the same amount of nucleation regardless of the shape of the parent crystal. This is due to the substantially greater effect of secondary nucleation over primary nucleation of the original crystal. This was proven through both temperature and shape dependent nuclei stimulated growth experiments to confirm that in cases of secondary nucleation only the degree and temperature of supercooling change the nucleation rate, whereas the parent crystal only serves to act as a catalytic initiator of the process.
Isotactic poly (vinylcyclohexane) (PVCH)	PVCH crystals were experimentally shown to increase their spread and lateral growth at high temperatures indicating that although they were feasibly unable to reach Regime III temperatures, extrapolations and hypotheses from the experiment point to confirmation of the expected behavior in each of the three regimes. Experiments concluded that additional growth mechanisms such as crystal twinning and twin boundary interactions can alter the traditional LH theory, but further research is needed to model each individual influence.

Zinc oxide (ZnO)	Zinc oxide crystals were proven to undergo secondary nucleation under an odd mixture of conditions including the addition of a diamine as well as surface etching. Overall, testing has shown that the morphology of the secondary crystals can fluctuate greatly depending on the amount of diamine added, due to its ability to deplete the substrate and hinder the growth prematurely.

Polymer Field Theory

A polymer field theory is a statistical field theory describing the statistical behavior of a neutral or charged polymer system. It can be derived by transforming the partition function from its standard many-dimensional integral representation over the particle degrees of freedom in a functional integral representation over an auxiliary field function, using either the Hubbard-Stratonovich transformation or the delta-functional transformation. Computer simulations based on polymer field theories have been shown to deliver useful results, for example to calculate the structures and properties of polymer solutions (Baeurle 2007, Schmid 1998), polymer melts (Schmid 1998, Matsen 2002, Fredrickson 2002) and thermoplastics (Baeurle 2006).

Canonical Ensemble

Particle Representation of the Canonical Partition Function

The standard continuum model of flexible polymers, introduced by Edwards (Edwards 1965), treats a solution composed of n linear monodisperse homopolymers as a system of coarse-grained polymers, in which the statistical mechanics of the chains is described by the continuous Gaussian thread model (Baeurle 2007) and the solvent is taken into account implicitly. The Gaussian thread model can be considered as the continuum limit of the discrete Gaussian chain model, in which the polymers are described as continuous, linearly elastic filaments. The canonical partition function of such a system, kept at an inverse temperature $\beta = 1/k_B T$ and confined in a volume V, can be expressed as

$$Z(n,V,\beta) = \frac{1}{n!(\lambda_T^3)^{nN}} \prod_{j=1}^{n} \int D\mathbf{r}_j \exp\left(-\beta\Phi_0[\mathbf{r}] - \beta\bar{\Phi}[\mathbf{r}]\right), \quad (1)$$

where $\bar{\Phi}[\mathbf{r}]$ is the potential of mean force given by,

$$\bar{\Phi}[\mathbf{r}] = \frac{N^2}{2} \sum_{j=1}^{n}\sum_{k=1}^{n} \int_0^1 ds \int_0^1 ds' \bar{\Phi}\left(\left|\mathbf{r}_j(s) - \mathbf{r}_k(s')\right|\right) - \frac{1}{2}nN\bar{\Phi}(0), \quad (2)$$

representing the solvent-mediated non-bonded interactions among the segments, while $\Phi_0[\mathbf{r}]$ represents the harmonic binding energy of the chains. The latter energy contribution can be formulated as

$$\Phi_0[\mathbf{r}] = \frac{3k_B T}{2Nb^2} \sum_{l=1}^{n} \int_0^1 ds \left|\frac{d\mathbf{r}_l(s)}{ds}\right|^2,$$

where b is the statistical segment length and N the polymerization index.

Field-theoretic Transformation

To derive the basic field-theoretic representation of the canonical partition function, one introduces in the following the segment density operator of the polymer system

$$\hat{\rho}(\mathbf{r}) = N \sum_{j=1}^{n} \int_0^1 ds \delta\left(\mathbf{r} - \mathbf{r}_j(s)\right).$$

Using this definition, one can rewrite Eq. (2) as

$$\bar{\Phi}[\mathbf{r}] = \frac{1}{2} \int d\mathbf{r} \int d\mathbf{r}' \hat{\rho}(\mathbf{r}) \bar{\Phi}(|\mathbf{r}-\mathbf{r}'|) \hat{\rho}(\mathbf{r}') - \frac{1}{2} nN\bar{\Phi}(0). \qquad (3)$$

Next, one converts the model into a field theory by making use of the Hubbard-Stratonovich transformation or delta-functional transformation

$$\int D\rho\, \delta[\rho - \hat{\rho}] F[\rho] = F[\hat{\rho}], \qquad (4)$$

where $F[\hat{\rho}]$ is a functional and $\delta[\rho\ \ \hat{\rho}]$ is the delta functional given by

$$\delta[\rho - \hat{\rho}] = \int Dw e^{i\int d\mathbf{r} w(\mathbf{r})[\rho(\mathbf{r}) - \hat{\rho}(\mathbf{r})]}, \qquad (5)$$

with $w(\mathbf{r}) = \sum_{\mathbf{G}} w(\mathbf{G}) \exp[i\mathbf{Gr}]$ representing the auxiliary field function. Here we note that, expanding the field function in a Fourier series, implies that periodic boundary conditions are applied in all directions and that the $\mathbf{G}$-vectors designate the reciprocal lattice vectors of the supercell.

Basic field-theoretic Representation of Canonical Partition Function

Using the Eqs. (3), (4) and (5), we can recast the canonical partition function in Eq. (1) in field-theoretic representation, which leads to

$$Z(n,V,\beta) = Z_0 \int Dw \exp\left[-\frac{1}{2\beta V^2} \int d\mathbf{r} d\mathbf{r}' w(\mathbf{r}) \bar{\Phi}^{-1}(\mathbf{r}-\mathbf{r}') w(\mathbf{r}')\right] Q^n[iw], \qquad (6)$$

where

$$Z_0 = \frac{1}{n!}\left(\frac{\exp\left(\beta/2N\bar{\Phi}(0)\right)Z'}{\lambda^{3N}(T)}\right)^n$$

can be interpreted as the partition function for an ideal gas of non-interacting polymers and

$$Z' = \int D\mathbf{R} \exp\left[-\beta U_0(\mathbf{R})\right] \qquad (7)$$

is the path integral of a free polymer in a zero field with elastic energy

$$U_0[\mathbf{R}] = \frac{k_B T}{4R_{g0}^2} \int_0^1 ds \left|\frac{d\mathbf{R}(s)}{ds}\right|^2.$$

In the latter equation the unperturbed radius of gyration of a chain $R_{g0} = \sqrt{Nb^2/(6)}$. Moreover, in Eq. (6) the partition function of a single polymer, subjected to the field $w(\mathbf{R})$, is given by

$$Q[iw] = \frac{\int D\mathbf{R} \exp\left[-\beta U_0[\mathbf{R}] - iN\int_0^1 ds\, w(\mathbf{R}(s))\right]}{\int D\mathbf{R} \exp\left[-\beta U_0[\mathbf{R}]\right]}. \qquad (8)$$

Grand Canonical Ensemble

Basic field-theoretic Representation of Grand Canonical Partition Function

To derive the grand canonical partition function, we use its standard thermodynamic relation to the canonical partition function, given by

$$\Xi(\mu,V,\beta) = \sum_{n=0}^{\infty} e^{\beta\mu n} Z(n,V,\beta),$$

where μ is the chemical potential and $Z(n,V,\beta)$ is given by Eq. (6). Performing the sum, this provides the field-theoretic representation of the grand canonical partition function,

$$\Xi(\xi,V,\beta) = \gamma_{\bar{\Phi}} \int Dw \exp[-S[w]],$$

where

$$S[w] = \frac{1}{2\beta V^2} \int d\mathbf{r}d\mathbf{r}' w(\mathbf{r})\bar{\Phi}^{-1}(\mathbf{r}-\mathbf{r}')w(\mathbf{r}') - \xi Q[iw]$$

is the grand canonical action with $Q[iw]$ defined by Eq. (8) and the constant

$$\gamma_{\bar{\Phi}} = \frac{1}{\sqrt{2}} \prod_{\mathbf{G}} \left(\frac{1}{\pi\beta\bar{\Phi}(\mathbf{G})} \right)^{1/2}.$$

Moreover, the parameter related to the chemical potential is given by

$$\xi = \frac{\exp\left(\beta\mu + \beta/2N\bar{\Phi}(0)\right)Z'}{\lambda^{3N}(T)},$$

where Z' is provided by Eq. (7).

Mean Field Approximation

A standard approximation strategy for polymer field theories is the mean field (MF) approximation, which consists in replacing the many-body interaction term in the action by a term where all bodies of the system interact with an average effective field. This approach reduces any multi-body problem into an effective one-body problem by assuming that the partition function integral of the model is dominated by a single field configuration. A major benefit of solving problems with the MF approximation, or its numerical implementation commonly referred to as the self-consistent field theory (SCFT), is that it often provides some useful insights into the properties and behavior of complex many-body systems at relatively low computational cost. Successful applications of this approximation strategy can be found for various systems of polymers and complex fluids, like e.g. strongly segregated block copolymers of high molecular weight, highly concentrated neutral polymer solutions or highly concentrated block polyelectrolyte (PE) solutions (Schmid 1998, Matsen 2002, Fredrickson 2002). There are, however, a multitude of cases for which SCFT provides inaccurate or even qualitatively incorrect results (Baeurle 2006a). These comprise neutral polymer or polyelectrolyte solutions in dilute and semidilute concentration regimes, block copolymers near their order-disorder transition, polymer blends near their phase transitions, etc. In such situations the partition function integral defining the field-theoretic model is not entirely dominated by a single MF configuration and field configurations far from it can make important contributions, which require the use of more sophisticated calculation techniques beyond the MF level of approximation.

Higher-order Corrections

One possibility to face the problem is to calculate higher-order corrections to the MF approximation. Tsonchev et al. developed such a strategy including leading (one-loop) order fluctuation corrections, which allowed to gain new insights into the physics of confined PE solutions (Tsonchev 1999). However, in situations where the MF approximation is bad many computationally demanding higher-order corrections to the integral are necessary to get the desired accuracy.

Renormalization Techniques

An alternative theoretical tool to cope with strong fluctuations problems occurring in field theories has been provided in the late 1940s by the concept of renormalization, which has originally been devised to calculate functional integrals arising in quantum field theories (QFT's). In QFT's a standard approximation strategy is to expand the functional integrals in a power series in the coupling constant using perturbation theory. Unfortunately, generally most of the expansion terms turn out to be infinite, rendering such calculations impracticable (Shirkov 2001). A way to remove the infinities from QFT's is to make use of the concept of renormalization (Baeurle 2007). It mainly consists in replacing the bare values of the coupling parameters, like e.g. electric charges or masses, by renormalized coupling parameters and requiring that the physical quantities do not change under this transformation, thereby leading to finite terms in the perturbation expansion. A simple physical picture of the procedure of renormalization can be drawn from the example of a classical electrical charge, Q, inserted into a polarizable medium, such as in an electrolyte solution. At a distance r from the charge due to polarization of the medium, its Coulomb field will effectively depend on a function $Q(r)$, i.e. the effective (renormalized) charge, instead of the bare electrical charge, Q. At the beginning of the 1970s, K.G. Wilson further pioneered the power of renormalization concepts by developing the formalism of renormalization group (RG) theory, to investigate critical phenomena of statistical systems (Wilson 1971).

Renormalization Group Theory

The RG theory makes use of a series of RG transformations, each of which consists of a coarse-graining step followed by a change of scale (Wilson 1974). In case of statistical-mechanical problems the steps are implemented by successively eliminating and rescaling the degrees of freedom in the partition sum or integral that defines the model under consideration. De Gennes used this strategy to establish an analogy between the behavior of the zero-component classical vector model of ferromagnetism near the phase transition and a self-avoiding random walk of a polymer chain of infinite length on a lattice, to calculate the polymer excluded volume exponents (de Gennes 1972). Adapting this concept to field-theoretic functional integrals, implies to study in a systematic way how a field theory model changes while eliminating and rescaling a certain number of degrees of freedom from the partition function integral (Wilson 1974).

Hartree Renormalization

An alternative approach is known as the *Hartree approximation* or *self-consistent one-loop approximation* (Amit 1984). It takes advantage of Gaussian fluctuation corrections to the 0th-order MF contribution, to renormalize the model parameters and extract in a self-consistent way the dominant length scale of the concentration fluctuations in critical concentration regimes.

Tadpole Renormalization

In a more recent work Efimov and Nogovitsin showed that an alternative renormalization technique originating from QFT, based on the concept of *tadpole renormalization*, can be a very effective approach for computing functional integrals arising in statistical mechanics of classical many-particle systems (Efimov 1996). They demonstrated that the main contributions to classical partition function integrals are provided by low-order tadpole-type Feynman diagrams, which account for divergent contributions due to particle self-interaction. The renormalization procedure performed in this approach effects on the self-interaction contribution of a charge (like e.g. an electron or an ion), resulting from the static polarization induced in the vacuum due to the presence of that charge (Baeurle 2007). As evidenced by Efimov and Ganbold in an earlier work (Efimov 1991), the procedure of tadpole renormalization can be employed very effectively to remove the divergences from the action of the basic field-theoretic representation of the partition function and leads to an alternative functional integral representation, called the Gaussian equivalent representation (GER). They showed that the procedure provides functional integrals with significantly ameliorated convergence properties for analytical perturbation calculations. In subsequent works Baeurle et al. developed effective low-cost approximation methods based on the tadpole renormalization procedure, which have shown to deliver useful results for prototypical polymer and PE solutions (Baeurle 2006a, Baeurle 2006b, Baeurle 2007a).

Numerical Simulation

Another possibility is to use Monte Carlo (MC) algorithms and to sample the full partition function integral in field-theoretic formulation. The resulting procedure is then called a polymer field-theoretic simulation. In a recent work, however, Baeurle demonstrated that MC sampling in conjunction with the basic field-theoretic representation is impracticable due to the so-called numerical sign problem (Baeurle 2002). The difficulty is related to the complex and oscillatory nature of the resulting distribution function, which causes a bad statistical convergence of the ensemble averages of the desired thermodynamic and structural quantities. In such cases special analytical and numerical techniques are necessary to accelerate the statistical convergence (Baeurle 2003, Baeurle 2003a, Baeurle 2004).

Mean Field Representation

To make the methodology amenable for computation, Baeurle proposed to shift the contour of integration of the partition function integral through the homogeneous MF solution using Cauchy's integral theorem, providing its so-called *mean-field representation*. This strategy was previously successfully employed by Baer et al. in field-theoretic electronic structure calculations (Baer 1998). Baeurle could demonstrate that this technique provides a significant acceleration of the statistical convergence of the ensemble averages in the MC sampling procedure (Baeurle 2002, Baeurle 2002a).

Gaussian Equivalent Representation

In subsequent works Baeurle et al. (Baeurle 2002, Baeurle 2002a, Baeurle 2003, Baeurle 2003a, Baeurle 2004) applied the concept of tadpole renormalization, leading to the *Gaussian equivalent representation*of the partition function integral, in conjunction with advanced MC techniques in the grand canonical ensemble. They could convincingly demonstrate that this strategy provides a further boost in the statistical convergence of the desired ensemble averages (Baeurle 2002).

Flory–Fox Equation

In polymer chemistry, the Flory–Fox equation is a simple empirical formula that relates molecular weight to the glass transition temperature of a polymer system. The equation was first proposed in 1950 by Paul J. Flory and Thomas G. Fox while at Cornell University. Their work on the subject overturned the previously held theory that the glass transition temperature was the temperature at which viscosity reached a maximum. Instead, they demonstrated that the glass transition temperature is the temperature at which the free space available for molecular motions achieved a minimum value. While its accuracy is usually limited to samples of narrow range molecular weight distributions, it serves as a good starting point for more complex structure-property relationships.

Overview

The Flory–Fox equation relates the number-average molecular weight, M_n, to the glass transition temperature, T_g, as shown below:

$$T_g = T_{g,\infty} - \frac{K}{M_n}$$

where $T_{g,\infty}$ is the maximum glass transition temperature that can be achieved at a theoretical infinite molecular weight and K is an empirical parameter that is related to the free volume present in the polymer sample. It is this concept of "free volume" that is observed by the Flory–Fox equation.

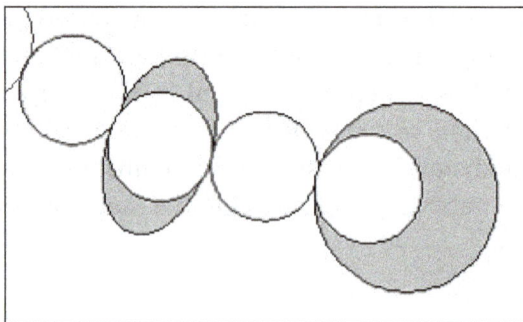

A polymer chain (represented by the white circles) exhibits more free volume (represented by the gray shading) at the ends of the chain than from units within the chain.

Free volume can be most easily understood as a polymer chain's "elbow room" in relation to the other polymer chains surrounding it. The more elbow room a chain has, the easier it is for the chain to move and achieve different physical conformations. Free volume decreases upon cooling from the rubbery state until the glass transition temperature at which point it reaches some critical minimum value and molecular rearrangement is effectively "frozen" out, so the polymer chains lack sufficient free volume to achieve different physical conformations. This ability to achieve different physical conformations is called segmental mobility.

Free volume not only depends on temperature, but also on the number of polymer chain ends present in the system. End chain units exhibit greater free volume than units within the chain because the covalent bonds that make up the polymer are shorter than the intermolecular nearest neighbor distances found at the end of the chain. In other words, end chain units are less dense than the covalently bonded interchain units. This means that a polymer sample with long chain lengths (high molecular weights) will have fewer chain ends per total units and less free volume than a polymer sample consisting of short chains. In short, chain ends can be viewed as an "impurity" when considering the packing of chains, and more impurity results in a lower T_g.

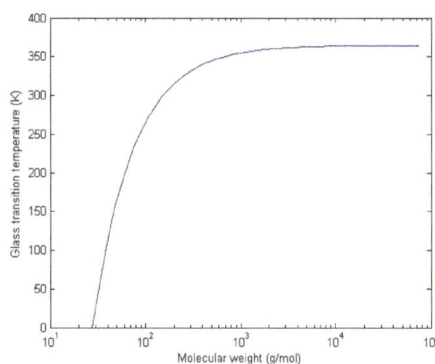

A plot showing the dependence of glass transition temperature (Kelvin) on molecular weight of the polymer chain (g/mol), as predicted by the Flory-Fox equation.

Thus, glass transition temperature is dependent on free volume, which in turn is dependent on the average molecular weight of the polymer sample. This relationship is described by the Flory–Fox equation. Low molecular weight values result in lower glass transition temperatures whereas increasing values of molecular weight result in an asymptotic approach of the glass transition temperature to $T_{g,\infty}$. The figure to the left clearly displays this relationship – as molecular weight increases, the glass transition temperature increases asymptotically to $T_{g,\infty}$ (in this arbitrary case shown in the image, $T_{g,\infty}$ = 365 K).

Alternative Equations

While the Flory–Fox equation describes many polymers very well, it is more reliable for large values of M_n and samples of narrow weight distribution. As a result, other equations have been proposed to provide better accuracy for certain polymers. For example:

$$T_g = T_{g,\infty} - \frac{K}{(M_n M_w)^{\frac{1}{2}}}.$$

This minor modification of the Flory–Fox equation, proposed by Ogawa, replaces the inverse dependence on M_n with the square of the product of the number-average molecular weight, M_n , and weight-average molecular weight, M_w . Additionally, the equation:

$$\frac{1}{T_g} = \frac{1}{T_{g,\infty}} + \frac{K}{T_{g,\infty}^2} \frac{1}{M_n}$$

was proposed by Fox and Loshaek, and has been applied to polystyrene, polymethylmethacrylate, and polyisobutylene, among others.

However, it is important to note that despite the dependence of T_g on molecular weight that the Flory-Fox and related equations describe, molecular weight is not necessarily a practical design parameter for controlling T_g because the range over which it can be changed without altering the physical properties of the polymer due to molecular weight change is small.

The Fox Equation

The Flory–Fox equation serves the purpose of providing a model for how glass transition temperature changes over a given molecular weight range. Another method to modify the glass transition temperature is to add a small amount of low molecular weight diluent, commonly known as a plasticizer, to the polymer. The presence of a low molecular weight additive increases the free volume of the system and subsequently lowers T_g, thus allowing for rubbery properties at lower temperatures. This effect is described by the Fox equation:

$$\frac{1}{T_g} = \frac{w_1}{T_{g,1}} + \frac{w_2}{T_{g,2}}.$$

Where w_1 and w_2 are weight fractions of components 1 and 2, respectively. In general, the accuracy of the Fox equation is very good and it is commonly also applied to predict the glass transition temperature in (miscible) polymer blends and statistical copolymers.

Mayo–Lewis Equation

The Mayo–Lewis equation or copolymer equation in polymer chemistry describes the distribution of monomers in a copolymer: It is named for Frank R. Mayo and Frederick M. Lewis.

Taking into consideration a monomer mix of two components M_1 and M_2 and the four different reactions that can take place at the reactive chain end terminating in either monomer (M^*) with their reaction rate constants k:

$$M_1^* + M_1 \xrightarrow{k_{11}} M_1 M_1^*$$

$$M_1^* + M_2 \xrightarrow{k_{12}} M_1 M_2^*$$

$$M_2^* + M_2 \xrightarrow{k_{22}} M_2 M_2^*$$

$$M_2^* + M_1 \xrightarrow{k_{21}} M_2 M_1^*$$

and with reactivity ratios defined as:

$$r_1 = \frac{k_{11}}{k_{12}}$$

$$r_2 = \frac{k_{22}}{k_{21}}$$

the copolymer equation is given as:

$$\frac{d[M_1]}{d[M_2]} = \frac{[M_1](r_1[M_1]+[M_2])}{[M_2]([M_1]+r_2[M_2])}$$

with the concentration of the components given in square brackets. The equation gives the copolymer composition at any instant during the polymerization.

Equation Derivation

Monomer 1 is consumed with reaction rate:

$$\frac{-d[M_1]}{dt} = k_{11}[M_1]\sum[M_1^*]+k_{21}[M_1]\sum[M_2^*]$$

with $\sum[M_x^*]$ the concentration of all the active centers terminating in monomer 1 or 2.

Likewise the rate of disappearance for monomer 2 is:

$$\frac{-d[M_2]}{dt} = k_{12}[M_2]\sum[M_1^*]+k_{22}[M_2]\sum[M_2^*]$$

Division of both equations yields:

$$\frac{d[M_1]}{d[M_2]} = \frac{[M_1]}{[M_2]}\left(\frac{k_{11}\dfrac{\sum[M_1^*]}{\sum[M_2^*]}+k_{21}}{k_{12}\dfrac{\sum[M_1^*]}{\sum[M_2^*]}+k_{22}}\right)$$

The ratio of active center concentrations can be found assuming steady state with:

$$\frac{d\sum[M_1^*]}{dt} = \frac{d\sum[M_2^*]}{dt} \approx 0$$

meaning that the concentration of active centres remains constant, the rate of formation for active center of monomer 1 is equal to the rate of their destruction or:

$$k_{21}[M_1]\sum[M_2^*] = k_{12}[M_2]\sum[M_1^*]$$

or

$$\frac{\sum[M_1^*]}{\sum[M_2^*]} = \frac{k_{21}[M_1]}{k_{12}[M_2]}$$

Substituting into the ratio of monomer consumption rates eliminates the radical concentrations and yields the Mayo-Lewis equation.

Instantaneous Form

It can often be useful to alter the copolymer equation by expressing concentration in terms of mole fractions. Mole fractions of monomers M_1 and M_2 in the feed are defined as f_1 and f_2 where

$$f_1 = 1 - f_2 = \frac{M_1}{(M_1 + M_2)}$$

Similarly, F represents the mole fraction of each monomer in the copolymer:

$$F_1 = 1 - F_2 = \frac{dM_1}{d(M_1 + M_2)}$$

These equations can be combined with the Mayo-Lewis Equation to give

$$F_1 = 1 - F_2 = \frac{r_1 f_1^2 + f_1 f_2}{r_1 f_1^2 + 2 f_1 f_2 + r_2 f_2^2}$$

This equation gives the instantaneous copolymer composition. It is important to note that the feed and copolymer compositions can change as polymerization proceeds.

Limiting Cases

Reactivity ratios indicate preference for propagation. Large r indicates a tendency for M_1^* to add M_1, while small values indicate a tendency for M_1^* to add M_2. From the definition of reactivity ratios, several special cases can be derived:

- $r_1 = r_2 \gg 1$ with both reactivity ratios very high the two monomers only react with themselves and not each other leading to a mixture of two homopolymers.

- $r_1 = r_2 > 1$ with both ratios larger than 1, homopolymerization of component M_1 is favored but in the event of a crosspolymerization by M_2 the chain-end will continue giving rise to block copolymer

- $r_1 = r_2 \approx 1$ with both ratios around 1, monomer 1 will react as fast with another monomer 1 or monomer 2 and a random copolymer is formed.

- $r_1 = r_2 \approx 0$ with both values approaching 0 the monomers are unable to homopolymerize and only add each other resulting in an alternating polymer

- $r_1 \gg 1 \gg r_2$ In the initial stage of the copolymerization monomer 1 is incorporated faster and the copolymer is rich in monomer 1. When this monomer gets depleted, more monomer 2 segments are added. This is called composition drift.

An example case is maleic anhydride and styrene, with reactivity ratios:

- Maleic anhydride (r_1 = 0.01) & styrene (r_2, = 0.02)

Neither of these compounds homopolymerize and instead they react together to give almost exclusively alternating copolymer.

When both $r < 1$, the system has an azeotrope, where feed and copolymer composition are the same.

Calculation of Reactivity Ratios

Calculation of reactivity ratios generally involves carrying out several polymerizations at varying monomer ratios. The copolymer composition can be analysed with methods such as Proton nuclear magnetic resonance, Carbon-13 nuclear magnetic resonance, or Fourier transform infrared spectroscopy. The polymerizations are also carried out at low conversions, so monomer concentrations can be assumed to be constant. With all the other parameters in the copolymer equation known, r_1 and r_2 can be found.

Curve Fitting

One of the simplest methods for finding reactivity ratios is plotting the copolymer equation and using least squares analysis to find the r_1, r_2 pair that gives the best fit curve.

Mayo-Lewis Method

The Mayo-Lewis method uses a form of the copolymer equation relating r_1 to r_2 :

$$r_2 = \frac{f_1}{f_2}\left[\frac{F_2}{F_1}(1+\frac{f_1 r_1}{f_2})-1\right]$$

For each different monomer composition, a line is generated using arbitrary r_1 values. The intersection of these lines is the r_1, r_2 for the system. More frequently, the lines do not intersect and the area in which most line intersect can be given as a range of r_1, and r_2 values.

Fineman-Ross Method

Fineman and Ross rearranged the copolymer equation into a linear form:

$$G = Hr_1 - r_2$$

where $G = \dfrac{f_1(2F_1-1)}{(1-f_1)F_1}$ and $H = \left[\dfrac{f_1^2(1-F_1)}{(1-f_1)^2 F_1}\right]..$

Thus, a plot of H versus G yields a straight line with slope r_1 and intercept $-r_2$

Kelen Tudos Method

The Fineman-Ross method can be biased towards points at low or high monomer concentration, so Kelen and Tudos introduced and arbitrary constant,

$$\alpha = (H_{min}H_{max})^{0.5}$$

where H_{min} and H_{max} are the highest and lowest values of H from the Fineman-Ross method. The data can be plotted in a linear form

$$\eta = \left[r_1 + \frac{r_2}{\alpha}\right]\mu - \frac{r_2}{\alpha}$$

where $\eta = G/(\alpha+H)$ and $\mu = H/(\alpha+H)$. Plotting η against μ yields a straight line that gives $-r_2/\alpha$ when $\mu = 0$ and r_1 when $\mu = 1$. This distributes the data more symmetrically and can yield better results.

Q-e Scheme

A semi-empirical method for the determination of reactivity ratios is called the Q-e scheme. This involves using two parameters for each monomer, Q and e. The reaction of M_1 radical with M_2 monomer is written as

$$k_{12} = P_1 Q_2 exp(-e_1 e_2)$$

while the reaction of M_1 radical with M_1 monomer is written as

$$k_{11} = P_1 Q_1 exp(-e_1 e_1)$$

Where Q is the measure of reactivity of monomer via resonance stabilization, and e is the measure of polarity of monomer (molecule or radical) via the effect of functional groups on vinyl groups. Using these definitions, r_1 and r_2 can be found by the ratio of the terms. An advantage of this system is that reactivity ratios can be found using tabulated Q-e values of monomers regardless or what the monomer pair is in the system.

References

- Burchard, W (1983). "Solution Thermodyanmics of Non-Ionic Water Soluble Polymers.". In Finch, C. Chemistry and Technology of Water-Soluble Polymers. Springer. pp. 125–142. ISBN 978-1-4757-9661-2.

- Franks, F (1983). "Water Solubility and Sensitivity-Hydration Effects.". In Finch, C. Chemistry and Technology of Water-Soluble Polymers. Springer. pp. 157–178. ISBN 978-1-4757-9661-2.

- Muthukumar, M (2004). "Nucleation in Polymer Crystallization". Advances in Chemical Physics. 128. ISBN 0-471-44528-2.

- Amit, D.J. (1984). "Field theory, the renormalization group, and critical phenomena". Singapore, World Scientific. ISBN 9812561196.

- Young, Robert J. (1983). Introduction to polymers ([Reprinted with additional material] ed.). London: Chapman and Hall. ISBN 0-412-22170-5.

- Fox, T.G.; Loshaek, S. (1955), "Influence of molecular weight and degree of crosslinking on the specific volume and glass temperature of polymers", Journal of Polymer Science, 371: 15, doi:10.1002/pol.1955.120158006

Polymers: An Overview

Large molecules that contain repeated subunits are polymers. A molecule that binds chemically to other molecules to form a polymer is known as a monomer. Polymers have a number of classes such as biopolymer, conductive polymer, copolymer, silicone and smart polymer. This chapter is an overview of the subject matter incorporating all the major aspects of polymer chemistry.

Polymer

A polymer is a molecule, made from joining together many small molecules called monomers. The word "polymer" can be broken down into "poly" (meaning "many" in Greek) and "mer" (meaning "unit"). This shows how the chemical composition of a polymer consists of many smaller units (monomers) bonded together into a larger molecule. A chemical reaction bonding monomers together to make a polymer is called polymerization.

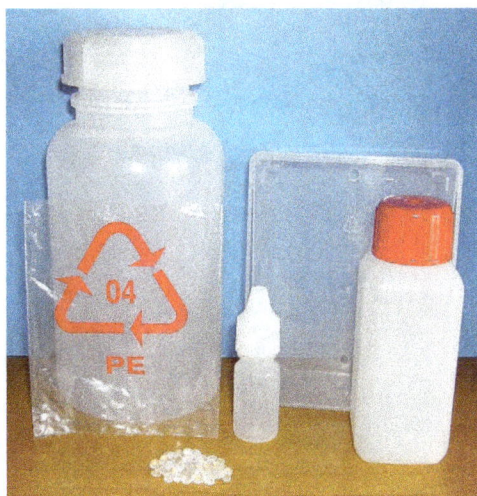

Objects made of polymers polyethylene and polypropylene

Some polymers are natural and made by organisms. Proteins have polypeptide molecules, which are natural polymers made from various amino acid monomer units. Nucleic acids are huge natural polymers made up of millions of nucleotide units. Cellulose and starch (two types of carbohydrate) are also natural polymers made up of glucopyranose monomer bonded together in different ways. Some polymers are man-made. Plastics, rubber, and fibers are made up of polymers.

If the "units" called monomers in a polymer are all the same, then the polymer is called a "homopolymer". Homopolymers are named by adding the prefix poly- before the monomer name from which the polymer is made. For example, a polymer made by bonding styrene monomer molecules together is called polystyrene.

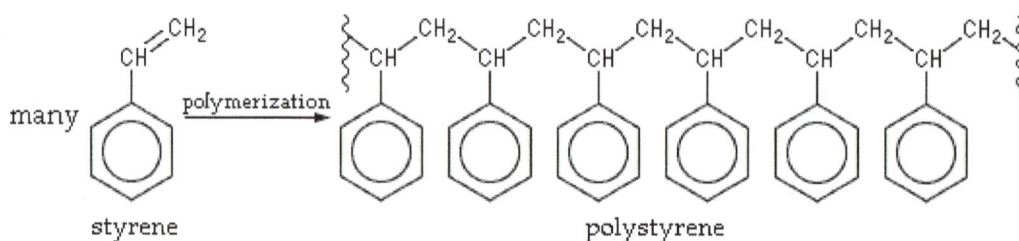

styrene   polystyrene

Many styrene molecules join together to make a polystyrene molecule. The squiggly lines at both ends of the polymer mean that just a short section of a long molecule is shown here.

If the monomers are not all the same, the polymer is called a "copolymer" or a "heteropolymer".

Many polymer molecules are like *chains* where the monomer units are the links. Polymer molecules can be straight-chain, have *branching* from the main chain, or *cross-linking* between chains. As an example of cross-linking, sulfhydryl (-S-H) groups in two cysteine amino acid units in polypeptide chains can bond together to make a disulfide bridge (-S-S-) joining the chains together.

Classes of Polymers

Biopolymer

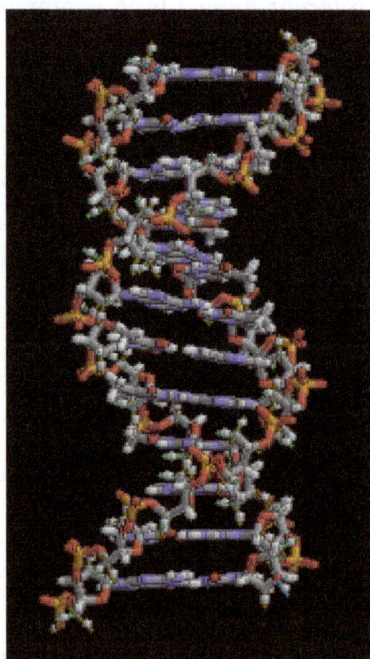

In the structure of DNA is a pair of **biopolymers**, polynucleotides, forming the double helix

Biopolymers are polymers produced by living organisms; in other words, they are polymeric biomolecules. Since they are polymers, biopolymers contain monomeric units that are covalently bonded to form larger structures. There are three main classes of biopolymers, classified according to the monomeric units used and the structure of the biopolymer formed: polynucleotides (RNA and DNA), which are long polymers composed of 13 or more nucleotide monomers; polypeptides, which are short polymers of amino acids; and polysaccharides, which are often linear bonded polymeric carbohydrate structures.

IUPAC Definition

Substance composed of one type of *biomacromolecules*.

Note 1: Modified from the definition given in ref. in order to avoid confusion between *polymer* and *macromolecule* in the fields of proteins, polysaccharides, polynucleotides, and bacterial aliphatic polyesters.

Note 2: The use of the term "biomacromolecule" is recommended when molecular characteristics are considered.

Cellulose is the most common organic compound and biopolymer on Earth. About 33 percent of all plant matter is cellulose. The cellulose content of cotton is 90 percent, for wood is 50 percent.

Biopolymers vs Synthetic Polymers

A major defining difference between biopolymers and other polymers can be found in their structures. All polymers are made of repetitive units called monomers. Biopolymers often have a well-defined structure, though this is not a defining characteristic (example: lignocellulose): The exact chemical composition and the sequence in which these units are arranged is called the primary structure, in the case of proteins. Many biopolymers spontaneously fold into characteristic compact shapes, which determine their biological functions and depend in a complicated way on their primary structures. Structural biology is the study of the structural properties of the biopolymers. In contrast, most synthetic polymers have much simpler and more random (or stochastic) structures. This fact leads to a molecular mass distribution that is missing in biopolymers. In fact, as their synthesis is controlled by a template-directed process in most *in vivo* systems, all biopolymers of a type (say one specific protein) are all alike: they all contain the similar sequences and numbers of monomers and thus all have the same mass. This phenomenon is called monodispersity in contrast to the polydispersity encountered in synthetic polymers. As a result, biopolymers have a polydispersity index of 1.

Conventions and Nomenclature

Polypeptides

The convention for a polypeptide is to list its constituent amino acid residues as they occur from the amino terminus to the carboxylic acid terminus. The amino acid residues are always joined by peptide bonds. Protein, though used colloquially to refer to any polypeptide, refers to larger or fully functional forms and can consist of several polypeptide chains as well as single chains. Proteins can also be modified to include non-peptide components, such as saccharide chains and lipids.

Nucleic Acids

The convention for a nucleic acid sequence is to list the nucleotides as they occur from the 5' end to the 3' end of the polymer chain, where 5' and 3' refer to the numbering of carbons around the ribose ring which participate in forming the phosphate diester linkages of the chain. Such a sequence is called the primary structure of the biopolymer.

Sugars

Sugar-based biopolymers are often difficult with regards to convention. Sugar polymers can be linear or branched and are typically joined with glycosidic bonds. The exact placement of the linkage can vary, and the orientation of the linking functional groups is also important, resulting in α- and β-glycosidic bonds with numbering definitive of the linking carbons' location in the ring. In addition, many saccharide units can undergo various chemical modifications, such as amination, and can even form parts of other molecules, such as glycoproteins.

Structural Characterization

There are a number of biophysical techniques for determining sequence information. Protein sequence can be determined by Edman degradation, in which the N-terminal residues are hydrolyzed from the chain one at a time, derivatized, and then identified. Mass spectrometer techniques can also be used. Nucleic acid sequence can be determined using gel electrophoresis and capillary electrophoresis. Lastly, mechanical properties of these biopolymers can often be measured using optical tweezers or atomic force microscopy. Dual polarization interferometry can be used to measure the conformational changes or self-assembly of these materials when stimulated by pH, temperature, ionic strength or other binding partners.

Biopolymers as Materials

Some biopolymers- such as (PLA), naturally occurring zein, and poly-3-hydroxybutyrate can be used as plastics, replacing the need for polystyrene or polyethylene based plastics.

Some plastics are now referred to as being 'degradable', 'oxy-degradable' or 'UV-degradable'. This means that they break down when exposed to light or air, but these plastics are still primarily (as much as 98 per cent) oil-based and are not currently certified as 'biodegradable' under the European Union directive on Packaging and Packaging Waste (94/62/EC). Biopolymers will break down, and some are suitable for domestic composting.

Biopolymers (also called renewable polymers) are produced from biomass for use in the packaging industry. Biomass comes from crops such as sugar beet, potatoes or wheat: when used to produce biopolymers, these are classified as non food crops. These can be converted in the following pathways:

Sugar beet > Glyconic acid > Polyglyconic acid

Starch > (fermentation) > Lactic acid > Polylactic acid (PLA)

Biomass > (fermentation) > Bioethanol > Ethene > Polyethylene

Many types of packaging can be made from biopolymers: food trays, blown starch pellets for shipping fragile goods, thin films for wrapping.

Environmental Impacts

Biopolymers can be sustainable, carbon neutral and are always renewable, because they are made from plant materials which can be grown indefinitely. These plant materials come from agricultural non food crops. Therefore, the use of biopolymers would create a sustainable industry. In contrast, the feedstocks for polymers derived from petrochemicals will eventually deplete. In addition, biopolymers have the potential to cut carbon emissions and reduce CO_2 quantities in the atmosphere: this is because the CO_2 released when they degrade can be reabsorbed by crops grown to replace them: this makes them close to carbon neutral.

Biopolymers are biodegradable, and some are also compostable. Some biopolymers are biodegradable: they are broken down into CO_2 and water by microorganisms. Some of these biodegradable biopolymers are compostable: they can be put into an industrial composting process and will break down by 90% within six months. Biopolymers that do this can be marked with a 'compostable' symbol, under European Standard EN 13432 (2000). Packaging marked with this symbol can be put into industrial composting processes and will break down within six months or less. An example of a compostable polymer is PLA film under 20μm thick: films which are thicker than that do not qualify as compostable, even though they are biodegradable. In Europe there is a home composting standard and associated logo that enables consumers to identify and dispose of packaging in their compost heap.

Conductive Polymer

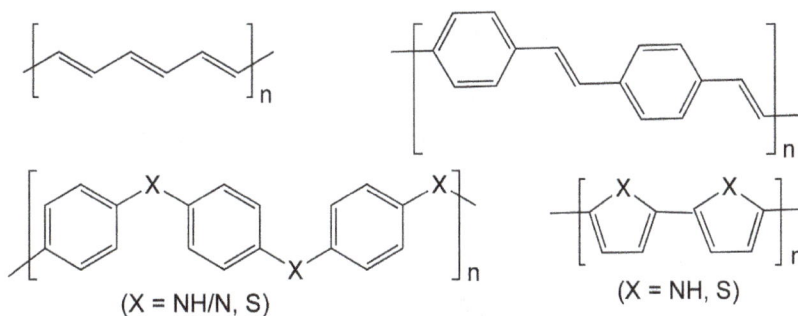

Chemical structures of some conductive polymers. From top left clockwise: polyacetylene; polyphenylene vinylene; polypyrrole (X = NH) and polythiophene (X = S); and polyaniline (X = NH/N) and polyphenylene sulfide (X = S).

Conductive polymers or, more precisely, intrinsically conducting polymers (ICPs) are organic polymers that conduct electricity. Such compounds may have metallic conductivity or can be semiconductors. The biggest advantage of conductive polymers is their processability, mainly by dispersion. Conductive polymers are generally not thermoplastics, *i.e.*, they are not thermoformable. But, like insulating polymers, they are organic materials. They can offer high electrical conductivity but do not show similar mechanical properties to other commercially available polymers. The electrical properties can be fine-tuned using the methods of organic synthesis and by advanced dispersion techniques.

History

Polyaniline was first described in the mid-19th century by Henry Letheby, who investigated the electrochemical and chemical oxidation products of aniline in acidic media. He noted that reduced form was colourless but the oxidized forms were deep blue.

The first highly-conductive organic compounds were the charge transfer complexes. In the 1950s, researchers reported that polycyclic aromatic compounds formed semi-conducting charge-transfer complex salts with halogens. In 1954, researchers at Bell Labs and elsewhere reported organic charge transfer complexes with resistivities as low as 8 ohms-cm. In the early 1970s, researchers demonstrated salts of tetrathiafulvalene show almost metallic conductivity, while superconductivity was demonstrated in 1980. Broad research on charge transfer salts continues today. While these compounds were technically not polymers, this indicated that organic compounds can carry current. While organic conductors were previously intermittently discussed, the field was particularly energized by the prediction of superconductivity following the discovery of BCS theory.

In 1963 Australians B.A. Bolto, D.E. Weiss, and coworkers reported derivatives of polypyrrole with resistivities as low as 1 ohm·cm. cites multiple reports of similar high-conductivity oxidized polyacetylenes. With the notable exception of charge transfer complexes (some of which are even superconductors), organic molecules were previously considered insulators or at best weakly conducting semiconductors. Subsequently, DeSurville and coworkers reported high conductivity in a polyaniline. Likewise, in 1980, Diaz and Logan reported films of polyaniline that can serve as electrodes.

While mostly operating in the quantum realm of less than 100 nanometers, "molecular" electronic processes can collectively manifest on a macro scale. Examples include quantum tunneling, negative resistance, phonon-assisted hopping and polarons. In 1977, Alan J. Heeger, Alan MacDiarmid and Hideki Shirakawa reported similar high conductivity in oxidized iodine-doped polyacetylene. For this research, they were awarded the 2000 Nobel Prize in Chemistry *"for the discovery and development of conductive polymers."* Polyacetylene itself did not find practical applications, but drew the attention of scientists and encouraged the rapid growth of the field. Since the late 1980s, organic light-emitting diodes (OLEDs) have emerged as an important application of conducting polymers.

Types

The linear-backbone "polymer blacks" (polyacetylene, polypyrrole, and polyaniline) and their copolymers are the main class of conductive polymers. Poly(p-phenylene vinylene) (PPV) and its soluble derivatives have emerged as the prototypical electroluminescent semiconducting polymers. Today, poly(3-alkylthiophenes) are the archetypical materials for solar cells and transistors.

The following table presents some organic conductive polymers according to their composition. The well-studied classes are written in bold and *the less well studied ones are in italic.*

The main chain contains	Heteroatoms present		
	No heteroatom	Nitrogen-containing	Sulfur-containing
Aromatic cycles	• *Poly(fluorene)s* • *polyphenylenes* • *polypyrenes* • *polyazulenes* • *polynaphthalenes*	The N is in the aromatic cycle: • **poly(pyrrole)s (PPY)** • *polycarbazoles* • *polyindoles* • *polyazepines* The N is outside the aromatic cycle: • **polyanilines (PANI)**	The S is in the aromatic cycle: • **poly(thiophene)s (PT)** • **poly(3,4-ethylenedioxythiophene) (PEDOT)** The S is outside the aromatic cycle: • **poly(p-phenylene sulfide) (PPS)**

Double bonds	• **Poly(acetylene)s (PAC)**		
Aromatic cycles and double bonds	• **Poly(p-phenylene vinylene) (PPV)**		

Synthesis

Conductive polymers are prepared by many methods. Most conductive polymers are prepared by oxidative coupling of monocyclic precursors. Such reactions entail dehydrogenation:

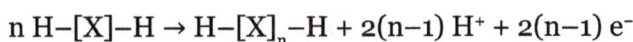

$$n \, H–[X]–H \rightarrow H–[X]_n–H + 2(n–1) \, H^+ + 2(n–1) \, e^-$$

The low solubility of most polymers presents challenges. Some researchers have addressed this through the formation of nanostructures and surfactant-stabilized conducting polymer dispersions in water. These include polyaniline nanofibers and PEDOT:PSS. These materials have lower molecular weights than that of some materials previously explored in the literature. However, in some cases, the molecular weight need not be high to achieve the desired properties.

Molecular Basis of Electrical Conductivity

The conductivity of such polymers is the result of several processes. For example, in traditional polymers such as polyethylenes, the valence electrons are bound in sp^3 hybridized covalent bonds. Such "sigma-bonding electrons" have low mobility and do not contribute to the electrical conductivity of the material. However, in conjugated materials, the situation is completely different. Conducting polymers have backbones of contiguous sp^2 hybridized carbon centers. One valence electron on each center resides in a p_z orbital, which is orthogonal to the other three sigma-bonds. All the p_z orbitals combine with each other to a molecule wide delocalized set of orbitals. The electrons in these delocalized orbitals have high mobility when the material is "doped" by oxidation, which removes some of these delocalized electrons. Thus, the conjugated p-orbitals form a one-dimensional electronic band, and the electrons within this band become mobile when it is partially emptied. The band structures of conductive polymers can easily be calculated with a tight binding model. In principle, these same materials can be doped by reduction, which adds electrons to an otherwise unfilled band. In practice, most organic conductors are doped oxidatively to give p-type materials. The redox doping of organic conductors is analogous to the doping of silicon semiconductors, whereby a small fraction silicon atoms are replaced by electron-rich, *e.g.*, phosphorus, or electron-poor, *e.g.*, boron, atoms to create n-type and p-type semiconductors, respectively.

Although typically "doping" conductive polymers involves oxidizing or reducing the material, conductive organic polymers associated with a protic solvent may also be "self-doped."

Undoped conjugated polymers state are semiconductors or insulators. In such compounds, the energy gap can be > 2 eV, which is too great for thermally activated conduction. Therefore, undoped conjugated polymers, such as polythiophenes, polyacetylenes only have a low electrical conductivity of around 10^{-10} to 10^{-8} S/cm. Even at a very low level of doping (< 1%), electrical conductivity increases several orders of magnitude up to values of around 0.1 S/cm. Subsequent doping of the conducting polymers will result in a saturation of the conductivity at values around 0.1–10 kS/cm for different polymers. Highest values reported up to now are for the conductivity of stretch oriented polyacetylene with confirmed values of about 80 kS/cm. Although the pi-electrons in polyactetylene are delocalized along the chain, pristine polyacetylene is not a metal. Polyacetylene has alternating single and double bonds which have lengths of 1.44 and 1.36 Å, respectively. Upon doping, the bond alteration is diminished in conductivity increases. Non-doping increases in conductivity can also be accomplished in a field effect transistor (organic FET or OFET) and by irradiation. Some materials also exhibit negative differential resistance and voltage-controlled "switching" analogous to that seen in inorganic amorphous semiconductors.

Despite intensive research, the relationship between morphology, chain structure and conductivity is still poorly understood. Generally, it is assumed that conductivity should be higher for the higher degree of crystallinity and better alignment of the chains, however this could not be confirmed for polyaniline and was only recently confirmed for PEDOT, which are largely amorphous.

Properties and Applications

Due to their poor processability, conductive polymers have few large-scale applications. They have promise in antistatic materials and they have been incorporated into commercial displays and batteries, but there have been limitations due to the manufacturing costs, material inconsistencies, toxicity, poor solubility in solvents, and inability to directly melt process. Literature suggests they are also promising in organic solar cells, printing electronic circuits, organic light-emitting diodes, actuators, electrochromism, supercapacitors, chemical sensors and biosensors, flexible transparent displays, electromagnetic shielding and possibly replacement for the popular transparent conductor indium tin oxide. Another use is for microwave-absorbent coatings, particularly radar-absorptive coatings on stealth aircraft. Conducting polymers are rapidly gaining attraction in new applications with increasingly processable materials with better electrical and physical properties and lower costs. The new nanostructured forms of conducting polymers particularly, augment this field with their higher surface area and better dispersability.

With the availability of stable and reproducible dispersions, PEDOT and polyaniline have gained some large scale applications. While PEDOT (poly(3,4-ethylenedioxythiophene)) is mainly used in antistatic applications and as a transparent conductive layer in form of PEDOT:PSS dispersions (PSS=polystyrene sulfonic acid), polyaniline is widely used for printed circuit board manufacturing – in the final finish, for protecting copper from corrosion and preventing its solderability.

Electroluminescence

Electroluminescence is light emission stimulated by electric current. In organic compounds, electroluminescence has been known since the early 1950s, when Bernanose and coworkers first produced electroluminescence in crystalline thin films of acridine orange and quinacrine. In 1960, researchers at Dow Chemical developed AC-driven electroluminescent cells using doping. In some cases, similar light emission is observed when a voltage is applied to a thin layer of a conductive organic polymer film. While electroluminescence was originally mostly of academic interest, the increased conductivity of modern conductive polymers means enough power can be put through the device at low voltages to generate practical amounts of light. This property has led to the development of flat panel displays using organic LEDs, solar panels, and optical amplifiers.

Barriers to Applications

Since most conductive polymers require oxidative doping, the properties of the resulting state are crucial. Such materials are salt-like (polymer salt), which diminishes their solubility in organic solvents and water and hence their processability. Furthermore, the charged organic backbone is often unstable towards atmospheric moisture. The poor processability for many polymers requires the introduction of solubilizing or substituents, which can further complicate the synthesis.

Experimental and theoretical thermodynamical evidence suggests that conductive polymers may even be completely and principally insoluble so that they can only be processed by dispersion.

Trends

Most recent emphasis is on organic light emitting diodes and organic polymer solar cells. The Organic Electronics Association is an international platform to promote applications of organic semiconductors. Conductive polymer products with embedded and improved electromagnetic interference (EMI) and electrostatic discharge (ESD) protection have led to both prototypes and products. For example, Polymer Electronics Research Center at University of Auckland is developing a range of novel DNA sensor technologies based on conducting polymers, photoluminescent polymers and inorganic nanocrystals (quantum dots) for simple, rapid and sensitive gene detection. Typical conductive polymers must be "doped" to produce high conductivity. As of 2001, there remains to be discovered an organic polymer that is *intrinsically* electrically conducting.

Copolymer

IUPAC Definition for Copolymer

A polymer derived from more than one species of monomer.

Note: Copolymers that are obtained by copolymerization of two monomer species are sometimes termed bipolymers, those obtained from three monomers terpolymers, those obtained from four monomers quaterpolymers, etc.

Alternating copolymers: A copolymer consisting of macromolecules comprising two species of monomeric units in alternating sequence.

Note: An alternating copolymer may be considered as a homopolymer derived from an implicit or hypothetical monomer.

Block copolymers: A portion of a macromolecule, comprising many constitutional units, that has at least one feature which is not present in the adjacent portions.

Graft macromolecule: A macromolecule with one or more species of block connected to the main chain as side-chains, these side-chains having constitutional or configurational features that differ from those in the main chain.

Vinyl Copolymer Milk

When two or more different monomers unite together to polymerize, their result is called a copolymer and its process is called copolymerization.

Commercially relevant copolymers include acrylonitrile butadiene styrene (ABS), styrene/butadiene co-polymer (SBR), nitrile rubber, styrene-acrylonitrile, styrene-isoprene-styrene (SIS) and ethylene-vinyl acetate.

Types of Copolymers

Different types of copolymers

Since a copolymer consists of at least two types of constituent units (also structural units), copolymers can be classified based on how these units are arranged along the chain. These include:

- Alternating copolymers with regular alternating A and B units (2)

- Periodic copolymers with A and B units arranged in a repeating sequence (e.g. (A-B-A-B-B-A-A-A-A-B-B-B)$_n$)

- Statistical copolymers are copolymers in which the sequence of monomer residues follows a statistical rule. If the probability of finding a given type monomer residue at a particular point in the chain is equal to the mole fraction of that monomer residue in the chain, then the polymer may be referred to as a truly random copolymer (3).

- Block copolymers comprise two or more homopolymer subunits linked by covalent bonds (4). The union of the homopolymer subunits may require an intermediate non-repeating subunit, known as a junction block. Block copolymers with two or three distinct blocks are called diblock copolymers and triblock copolymers, respectively.

Copolymers may also be described in terms of the existence of or arrangement of branches in the polymer structure. Linear copolymers consist of a single main chain whereas branched copolymers consist of a single main chain with one or more polymeric side chains.

Other special types of branched copolymers include star copolymers, brush copolymers, and comb copolymers. In gradient copolymers the monomer composition changes gradually along the chain.

A terpolymer is a copolymer consisting of three distinct monomers. The term is derived from *ter* (Latin), meaning thrice, and polymer.

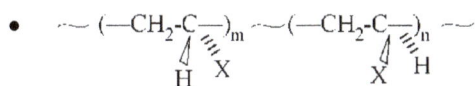

- $\sim (-CH_2\text{-}C-)_m \sim (-CH_2\text{-}C-)_n \sim$
  $\quad\quad H\ X \quad\quad\quad X\ H$

A special structure can be formed from one monomer where now the distinguishing feature is the tacticity of each block.

Graft Copolymers

Graft copolymers are a special type of branched copolymer in which the side chains are structurally distinct from the main chain. The illustration (5) depicts a special case where the main chain and side chains are composed of distinct homopolymers. However, the individual chains of a graft copolymer may be homopolymers or copolymers. Note that different copolymer sequencing is sufficient to define a structural difference, thus an A-B diblock copolymer with A-B alternating copolymer side chains is properly called a graft copolymer.

For example, suppose we perform a free-radical polymerization of styrene in the presence of polybutadiene, a synthetic rubber, which retains one reactive C=C double bond per residue. We get polystyrene chains growing out in either direction from some of the places where there were double bonds, with a one-carbon rearrangement. Or to look at it the other way around, the result is a polystyrene backbone with polybutadiene chains growing out of it in both directions. This is an interesting copolymer variant in that one of the ingredients was a polymer to begin with.

As with block copolymers, the quasi-composite product has properties of both "components". In the example cited, the rubbery chains absorb energy when the substance is hit, so it is much less brittle than ordinary polystyrene. The product is called high-impact polystyrene, or HIPS.

Block Copolymers

One kind of copolymer is called a "block copolymer". Block copolymers are made up of blocks of different polymerized monomers and is usually made by first polymerizing styrene, and then subsequently polymerizing methyl methacrylate (MMA) from the reactive end of the polystyrene chains. This polymer is a "diblock copolymer" because it contains two different chemical blocks. Triblocks, tetrablocks, multiblocks, etc. can also be made. Diblock copolymers are made using living polymerization techniques, such as atom transfer free radical polymerization (ATRP), reversible addition fragmentation chain transfer (RAFT), ring-opening metathesis polymerization (ROMP), and living cationic or living anionic polymerizations. An emerging technique is chain shuttling polymerization.

The "blockiness" of a copolymer is a measure of the adjacency of comonomers vs their statistical distribution. Many or even most synthetic polymers are in fact copolymers, containing about 1-20% of a minority monomoner. In such cases, blockiness is undesirable.

Phase Separation

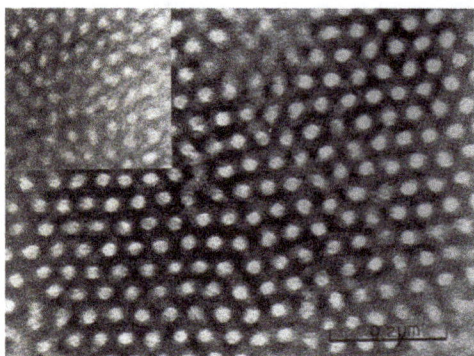

SBS block copolymer in TEM

Block copolymers are interesting because they can "microphase separate" to form periodic nano-structures, as in the styrene-butadiene-styrene block copolymer shown at right. The polymer is known as Kraton and is used for shoe soles and adhesives. Owing to the microfine structure, the transmission electron microscope or TEM was needed to examine the structure. The butadiene matrix was stained with osmium tetroxide to provide contrast in the image. The material was made by living polymerization so that the blocks are almost monodisperse, so helping to create a very regular microstructure. The molecular weight of the polystyrene blocks in the main picture is 102,000; the inset picture has a molecular weight of 91,000, producing slightly smaller domains.

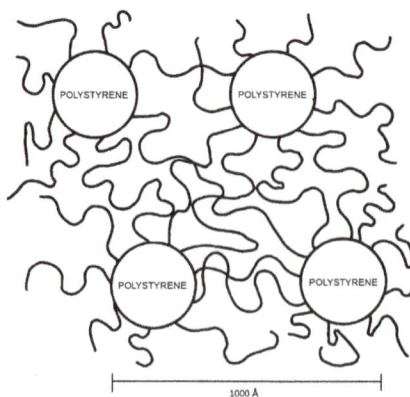

SBS block copolymer schematic microstructure

Microphase separation is a situation similar to that of oil and water. Oil and water are immiscible - they phase separate. Due to incompatibility between the blocks, block copolymers undergo a similar phase separation. Because the blocks are covalently bonded to each other, they cannot demix macroscopically as water and oil. In "microphase separation" the blocks form nanometer-sized structures. Depending on the relative lengths of each block, several morphologies can be obtained. In diblock copolymers, sufficiently different block lengths lead to nanometer-sized spheres of one block in a matrix of the second (for example PMMA in polystyrene). Using less different block lengths, a "hexagonally packed cylinder" geometry can be obtained. Blocks of similar length form layers (often called lamellae in the technical literature). Between the cylindrical and lamellar phase is the gyroid phase. The nanoscale structures created from block copolymers could potentially be used for creating devices for use in computer memory, nanoscale-templating and nanoscale separations.

Polymer scientists use thermodynamics to describe how the different blocks interact. The product of the degree of polymerization, n, and the Flory-Huggins interaction parameter, χ, gives an indication of how incompatible the two blocks are and whether or not they will microphase separate. For example, a diblock copolymer of symmetric composition will microphase separate if the product χN is greater than 10.5. If χN is less than 10.5, the blocks will mix and microphase separation is not observed. The incompatibility between the blocks also affects the solution behavior of these copolymers and their adsorption behavior on various surfaces.

Copolymer Equation

An alternating copolymer has the formula: -A-B-A-B-A-B-A-B-A-B-, or -(-A-B-)$_n$-. The molar ratios of the monomer in the polymer is close to one, which happens when the reactivity ratios r_1 & r_2 are close to zero, as given by the Mayo–Lewis equation also called the copolymerization equation:

$$\frac{d[M_1]}{d[M_2]} = \frac{[M_1]\left(r_1[M_1]+[M_2]\right)}{[M_2]\left([M_1]+r_2[M_2]\right)}$$

where $r_1 = k_{11}/k_{12}$ & $r_2 = k_{22}/k_{21}$

Copolymer Engineering

Copolymerization is used to modify the properties of manufactured plastics to meet specific needs, for example to reduce crystallinity, modify glass transition temperature or to improve solubility. It is a way of improving mechanical properties, in a technique known as rubber toughening. Elastomeric phases within a rigid matrix act as crack arrestors, and so increase the energy absorption when the material is impacted for example. Acrylonitrile butadiene styrene is a common example.

Polyester

Polyester is a category of polymers that contain the ester functional group in their main chain. As a specific material, it most commonly refers to a type called polyethylene terephthalate (PET). Polyesters include naturally occurring chemicals, such as in the cutin of plant cuticles, as well as synthetics through step-growth polymerization such as polybutyrate. Natural polyesters and a few synthetic ones are biodegradable, but most synthetic polyesters are not. This material is used very widely in clothing.

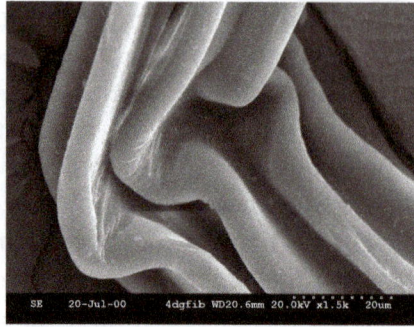

SEM picture of a bend in a high-surface area **polyester** fiber with a seven-lobed cross section

Close-up of a polyester shirt

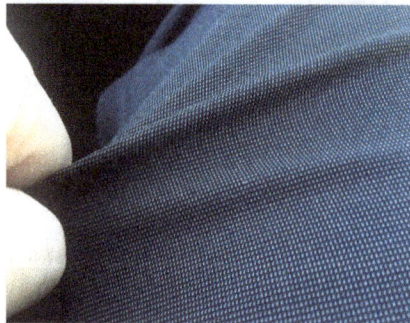

Stretching polyester fabric

Depending on the chemical structure, polyester can be a thermoplastic or thermoset. There are also polyester resins cured by hardeners; however, the most common polyesters are thermoplastics.

Fabrics woven or knitted from polyester thread or yarn are used extensively in apparel and home furnishings, from shirts and pants to jackets and hats, bed sheets, blankets, upholstered furniture and computer mouse mats. Industrial polyester fibers, yarns and ropes are used in tyre reinforcements, fabrics for conveyor belts, safety belts, coated fabrics and plastic reinforcements with high-energy absorption. Polyester fiber is used as cushioning and insulating material in pillows, comforters and upholstery padding. Polyester fabrics are highly stain-resistant— in fact, the only class of dyes which *can* be used to alter the color of polyester fabric are what are known as disperse dyes.

Polyester fibers are sometimes spun together with natural fibers to produce a cloth with blended properties. Cotton-polyester blends (polycotton) can be strong, wrinkle and tear-resistant, and reduce shrinking. Synthetic fibers in polyester also create materials with water, wind and environmental resistance compared to plant-derived fibers. Cons of cotton and polyester blends include being less breathable than cotton and trapping more moisture while sticking to the skin. They are also less fire resistant and can melt when ignited.

Polyester blends have been renamed so as to suggest their similarity or even superiority to natural fibers (for example, China silk, which is a term in the textiles industry for a 100% polyester fiber woven to resemble the sheen and durability of insect-derived silk).

Polyesters are also used to make bottles, films, tarpaulin, canoes, liquid crystal displays, holograms, filters, dielectric film for capacitors, film insulation for wire and insulating tapes. Polyesters are widely used as a finish on high-quality wood products such as guitars, pianos and vehicle/yacht interiors. Thixotropic properties of spray-applicable polyesters make them ideal for use on open-grain timbers, as they can quickly fill wood grain, with a high-build film thickness per coat. Cured polyesters can be sanded and polished to a high-gloss, durable finish.

Liquid crystalline polyesters are among the first industrially used liquid crystal polymers. They are used for their mechanical properties and heat-resistance. These traits are also important in their application as an abradable seal in jet engines.

Types

Polyesters as thermoplastics may change shape after the application of heat. While combustible at high temperatures, polyesters tend to shrink away from flames and self-extinguish upon ignition. Polyester fibers have high tenacity and E-modulus as well as low water absorption and minimal shrinkage in comparison with other industrial fibers.

Unsaturated polyesters (UPR) are thermosetting resins. They are used as casting materials, fiberglass laminating resins and non-metallic auto-body fillers. Fiberglass-reinforced unsaturated polyesters find wide application in bodies of yachts and as body parts of cars.

According to the composition of their main chain, polyesters can be:

Main chain composition	Type	Examples of	
		Polyesters	Manufacturing methods
Aliphatic	Homopolymer	Polyglycolide or polyglycolic acid (PGA)	Polycondensation of glycolic acid
		Polylactic acid (PLA)	Ring-opening polymerization of lactide
		Polycaprolactone (PCL)	Ring-opening polymerization of caprolactone
		Polyhydroxyalkanoate (PHA)	
		Polyhydroxybutyrate (PHB)	
	Copolymer	Polyethylene adipate (PEA)	
		Polybutylene succinate (PBS)	Polycondensation of succinic acid with 1,4-butanediol
		Poly(3-hydroxybutyrate-co-3-hydroxyvalerate) (PHBV)	Copolymerization of 3-hydroxybutanoic acid and 3-hydroxypentanoic acid, butyrolactone, and valerolactone (oligomeric aluminoxane as a catalyst)

Semi-aromatic	Copolymer	Polyethylene terephthalate (PET)	Polycondensation of terephthalic acid with ethylene glycol
		Polybutylene terephthalate (PBT)	Polycondensation of terephthalic acid with 1,4-butanediol
		Polytrimethylene terephthalate (PTT)	Polycondensation of terephthalic acid with 1,3-propanediol
		Polyethylene naphthalate (PEN)	Polycondensation of at least one naphthalene dicarboxylic acid with ethylene glycol
Aromatic	Copolymer	Vectran	Polycondensation of 4-hydroxybenzoic acid and 6-hydroxynaphthalene-2-carboxylic acid

Increasing the aromatic parts of polyesters increases their glass transition temperature, melting temperature, thermal stability, chemical stability...

Polyesters can also be telechelic oligomers like the polycaprolactone diol (PCL) and the polyethylene adipate diol (PEA). They are then used as prepolymers.

Industry

Basics

Polyester is a synthetic polymer made of purified terephthalic acid (PTA) or its dimethyl ester dimethyl terephthalate (DMT) and monoethylene glycol (MEG). With 18% market share of all plastic materials produced, it ranges third after polyethylene (33.5%) and polypropylene (19.5%).

The main raw materials are described as follows:

Purified terephthalic acid (PTA) CAS-No.: 100-21-0

> Synonym: 1,4 benzenedicarboxylic acid,
> Sum formula: $C_6H_4(COOH)_2$, mol. weight: 166.13

Dimethylterephthalate (DMT) CAS-No.: 120-61-6

> Synonym: 1,4 benzenedicarboxylic acid dimethyl ester,
> Sum formula: $C_6H_4(COOCH_3)_2$, mol. weight: 194.19

Mono-ethylene glycol (MEG) CAS No.: 107-21-1

> Synonym: 1,2 ethanediol,
> Sum formula: $C_2H_6O_2$, mol. weight: 62.07

To make a polymer of high molecular weight a catalyst is needed. The most common catalyst is antimony trioxide (or antimony tri-acetate):

Antimony trioxide (ATO) CAS-No.: 1309-64-4

> mol. weight: 291.51,
> Sum formula: Sb_2O_3

In 2008, about 10,000 tonnes Sb_2O_3 were used to produce around 49 million tonnes polyethylene terephthalate.

Polyester is described as follows:

Polyethylene terephthalate CAS-No.: 25038-59-9

> Synonyms/abbreviations: polyester, PET, PES,
> Sum formula: H-[$C_{10}H_8O_4$]-n=60–120 OH, mol. unit weight: 192.17

There are several reasons for the importance of polyester:

- The relatively easy accessible raw materials PTA or DMT and MEG
- The very well understood and described simple chemical process of polyester synthesis
- The low toxicity level of all raw materials and side products during polyester production and processing
- The possibility to produce PET in a closed loop at low emissions to the environment
- The outstanding mechanical and chemical properties of polyester
- The recyclability
- The wide variety of intermediate and final products made of polyester.

In the following table, the estimated world polyester production is shown. Main applications are textile polyester, bottle polyester resin, film polyester mainly for packaging and specialty polyesters for engineering plastics. According to this table, the world's total polyester production might exceed 50 million tons per annum before the year 2010.

World polyester production by year		
Product type	2002 (million tonnes/year)	2008 (million tonnes/year)
Textile-PET	20	39
Resin, bottle/A-PET	9	16
Film-PET	1.2	1.5
Special polyester	1	2.5
Total	**31.2**	**59**

Raw Material Producer

Polyester Processing

After the first stage of polymer production in the melt phase, the product stream divides into two different application areas which are mainly textile applications and packaging applications. In the following table, the main applications of textile and packaging of polyester are listed.

Textile and packaging polyester application list (melt or pellet)	
Textile	**Packaging**
Staple fiber (PSF)	Bottles for CSD, water, beer, juice, detergents, etc.
Filaments POY, DTY, FDY	A-PET film
Technical yarn and tire cord	Thermoforming
Non-woven and spunbond	biaxial-oriented film (BO-PET)
Mono-filament	Strapping

Abbreviations:

PSF

Polyester-staple fiber;

POY

Partially oriented yarn;

DTY

Draw-textured yarn;

FDY

Fully drawn yarn;

CSD

Carbonated soft drink;

A-PET

Amorphous polyester film;

BO-PET

Biaxial-oriented polyester film;

A comparable small market segment (much less than 1 million tonnes/year) of polyester is used to produce engineering plastics and masterbatch.

In order to produce the polyester melt with a high efficiency, high-output processing steps like staple fiber (50–300 tonnes/day per spinning line) or POY /FDY (up to 600 tonnes/day split into about 10 spinning machines) are meanwhile more and more vertically integrated direct processes. This means the polymer melt is directly converted into the textile fibers or filaments without the common step of pelletizing. We are talking about full vertical integration when polyester is produced at one site starting from crude oil or distillation products in the chain oil → benzene → PX → PTA → PET melt → fiber/filament or bottle-grade resin. Such integrated processes are meanwhile established in more or less interrupted processes at one production site. Eastman Chemicals were the first to introduce the idea of closing the chain from PX to PET resin with their so-called IN-TEGREX process. The capacity of such vertically integrated production sites is >1000 tonnes/day and can easily reach 2500 tonnes/day.

Besides the above-mentioned large processing units to produce staple fiber or yarns, there are ten thousands of small and very small processing plants, so that one can estimate that polyester is processed and recycled in more than 10 000 plants around the globe. This is without counting all the companies involved in the supply industry, beginning with engineering and processing machines and ending with special additives, stabilizers and colors. This is a gigantic industry complex and it is still growing by 4–8% per year, depending on the world region.

Synthesis

Synthesis of polyesters is generally achieved by a polycondensation reaction. See "condensation reactions in polymer chemistry". The general equation for the reaction of a diol with a diacid is :

$$(n+1) \ R(OH)_2 + n \ R'(COOH)_2 \rightarrow HO[ROOCR'COO]_nROH + 2n \ H_2O$$

Azeotrope Esterification

In this classical method, an alcohol and a carboxylic acid react to form a carboxylic ester. To assemble a polymer, the water formed by the reaction must be continually removed by azeotrope distillation.

Alcoholic Transesterification

Transesterification: An alcohol-terminated oligomer and an ester-terminated oligomer condense to form an ester linkage, with loss of an alcohol. R and R' are the two oligomer chains, R" is a sacrificial unit such as a methyl group (methanol is the byproduct of the esterification reaction).

Acylation (HCl Method)

The acid begins as an acid chloride, and thus the polycondensation proceeds with emission of hydrochloric acid (HCl) instead of water. This method can be carried out in solution or as an enamel.

Silyl Method

In this variant of the HCl method, the carboxylic acid chloride is converted with the trimethyl silyl ether of the alcohol component and production of trimethyl silyl chloride is obtained

Acetate Method (Esterification)

Silyl Acetate Method

Ring-opening Polymerization

Aliphatic polyesters can be assembled from lactones under very mild conditions, catalyzed anionically, cationically or metallorganically. A number of catalytic methods for the copolymerization of epoxides with cyclic anhydrides have also recently been shown to provide a wide array of functionalized polyesters, both saturated and unsaturated.

Biodegradation

The futuro house was made of fibreglass-reinforced polyester plastic; polyester-polyurethane, and poly(methylmethacrylate) one of them was found to be degrading by Cyanobacteria and Archaea.

Cross-linking

Unsaturated polyesters are thermosetting resins. They are generally copolymers prepared by polymerizing one or more diol with saturated and unsaturated dicarboxylic acids (maleic acid, fumaric acid...) or their anhydrides. The double bond of unsaturated polyesters reacts with a vinyl monomer, usually styrene, resulting in a 3-D cross-linked structure. This structure acts as a thermoset. The exothermic cross-linking reaction is initiated through a catalyst, usually an organic peroxide such as methyl ethyl ketone peroxide or benzoyl peroxide.

Environmental Concerns

Pollution of Freshwater and Seawater Habitats

A team at Plymouth University in the UK spent 12 months analysing what happened when a number of synthetic materials were washed at different temperatures in domestic washing machines, using different combinations of detergents, to quantify the microfibres shed. They found that an average washing load of 6 kg could release an estimated 137,951 fibres from polyester-cotton blend fabric, 496,030 fibres from polyester and 728,789 from acrylic.

Silicone

Silicone caulk can be used as a basic sealant against water and air penetration.

Silicones, also known as polysiloxanes, are polymers that include any inert, synthetic compound made up of repeating units of siloxane, which is a chain of alternating silicon atoms and oxygen atoms, frequently combined with carbon and/or hydrogen. They are typically heat-resistant and rubber-like, and are used in sealants, adhesives, lubricants, medicine, cooking utensils, and thermal and electrical insulation. Some common forms include silicone oil, silicone grease, silicone rubber, silicone resin, and silicone caulk.

Chemistry

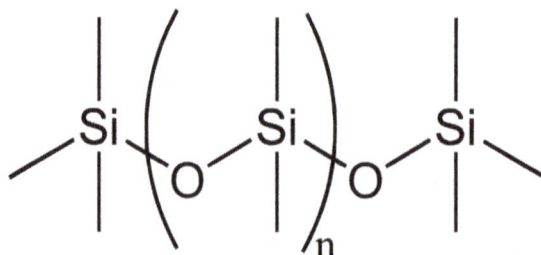

Chemical structure of the silicone polydimethylsiloxane (PDMS).

More precisely called polymerized siloxanes or polysiloxanes, silicones consist of an inorganic silicon-oxygen backbone chain (····-Si-O-Si-O-Si-O-····) with organic side groups attached to the silicon atoms. These silicon atoms are tetravalent. So, silicones are polymers constructed from inorganic-organic monomers. Silicones have in general the chemical formula $[R_2SiO]_n$, where R is an organic group such as methyl, ethyl, or phenyl.

In some cases, organic side groups can be used to link two or more of these -Si-O- backbones together. By varying the -Si-O- chain lengths, side groups, and crosslinking, silicones can be synthesized with a wide variety of properties and compositions. They can vary in consistency from liquid to gel to rubber to hard plastic. The most common siloxane is linear polydimethylsiloxane (PDMS), a silicone oil. The second largest group of silicone materials is based on silicone resins, which are formed by branched and cage-like oligosiloxanes.

Terminology and History

F. S. Kipping and Matt Saunders coined the word *silicone* in 1901 to describe polydiphenylsiloxane by analogy of its formula, Ph_2SiO (Ph stands for phenyl, C_6H_5), with the formula of the ketone benzophenone, Ph_2CO (his term was originally *silicoketone*). Kipping was well aware that polydiphenylsiloxane is polymeric whereas benzophenone is monomeric and noted that Ph_2SiO and Ph_2CO had very different chemistry. The discovery of the structural differences between Kippings' molecules and the ketones means that *silicone* is no longer the correct term (though it remains in common usage) and that the term *siloxanes* is correct according to the nomenclature of modern chemistry.

Silicone is sometimes mistakenly referred to as silicon. The chemical element silicon is a crystalline metalloid widely used in computers and other electronic equipment. Although silicones contain silicon atoms, they also include carbon, hydrogen, oxygen, and perhaps other kinds of atoms as well, and have physical and chemical properties that are very different from elemental silicon.

A true *silicone group* with a double bond between oxygen and silicon does not commonly exist in nature; chemists find that the silicon atom usually forms single bonds with each of two oxygen atoms, rather than a double bond to a single atom. Polysiloxanes are among the many substances commonly known as "silicones".

Molecules containing silicon-oxygen double bonds do exist and are called silanones but they are very reactive. Despite this, silanones are important as intermediates in gas-phase processes such as chemical vapor deposition in microelectronics production, and in the formation of ceramics by combustion.

Synthesis

Most common are materials based on polydimethylsiloxane, which is derived by hydrolysis of dimethyldichlorosilane. This dichloride reacts with water as follows:

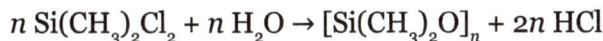

$$n\ Si(CH_3)_2Cl_2 + n\ H_2O \rightarrow [Si(CH_3)_2O]_n + 2n\ HCl$$

The polymerization typically produces linear chains capped with Si-Cl or Si-OH (silanol) groups. Under different conditions the polymer is a cyclic, not a chain.

For consumer applications such as caulks silyl acetates are used instead of silyl chlorides. The hydrolysis of the acetates produce the less dangerous acetic acid (the acid found in vinegar) as the reaction product of a much slower curing process. This chemistry is used in many consumer applications, such as silicone caulk and adhesives.

Branches or cross-links in the polymer chain can be introduced by using organosilicon precursors with fewer methyl groups, such as methyltrichlorosilane and methyltrimethoxysilane. Ideally, each molecule of such a compound becomes a branch point. This process can be used to produce hard silicone resins. Similarly, precursors with three methyl groups can be used to limit molecular weight, since each such molecule has only one reactive site and so forms the end of a siloxane chain.

Combustion

When silicone is burned in air or oxygen, it forms solid silica (silicon dioxide) as a white powder, char, and various gases. The readily dispersed powder is sometimes called silica fume.

Properties

Silicones exhibit many useful characteristics, including:

- Low thermal conductivity

- Low chemical reactivity

- Low toxicity

- Thermal stability (constancy of properties over a wide temperature range of −100 to 250 °C).

- The ability to repel water and form watertight seals.

- Does not stick to many substrates, but adheres very well to others, e.g. glass.

- Does not support microbiological growth.

- Resistance to oxygen, ozone, and ultraviolet (UV) light. This property has led to widespread use of silicones in the construction industry (e.g. coatings, fire protection, glazing seals) and the automotive industry (external gaskets, external trim).

- Electrical insulation properties. Because silicone can be formulated to be electrically insulative or conductive, it is suitable for a wide range of electrical applications.

- High gas permeability: at room temperature (25 °C), the permeability of silicone rubber for such gases as oxygen is approximately 400 times that of butyl rubber, making silicone useful for medical applications in which increased aeration is desired. Conversely, silicone rubbers cannot be used where gas-tight seals are necessary.

Uses

Silicones are used in many products. Ullmann's Encyclopedia of Industrial Chemistry lists the following major categories of application: Electrical (e.g., insulation), electronics (e.g., coatings), household (e.g., sealants for cooking apparatus), automobile (e.g., gaskets), aeroplane (e.g., seals), office machines (e.g., keyboard pads), medicine/dentistry (e.g., teeth impression molds), textiles/ paper (e.g., coatings). For these applications, an estimated 400,000 tons of silicones were produced in 1991. Specific examples, both large and small are presented below.

Automotive

In the automotive field, silicone grease is typically used as a lubricant for brake components since it is stable at high temperatures, is not water-soluble, and is far less likely than other lubricants to foul. It is also used as DOT 5 brake fluid.

Automotive spark plug wires are insulated by multiple layers of silicone to prevent sparks from jumping to adjacent wires, causing misfires. Silicone tubing is sometimes used in automotive intake systems (especially for engines with forced induction).

Sheet silicone is used to manufacture gaskets used in automotive engines, transmissions, and other applications.

Automotive body manufacturing plants and paint shops avoid silicones, as they may cause "fish eyes", small, circular craters in the finish.

Additionally, silicone compounds such as silicone rubber are used as coatings and sealants for airbags; the high strength of silicone rubber makes it an optimal adhesive/sealant for high impact airbags. Recent technological advancements allow convenient use of silicone in combination with thermoplastics to provide improvements in scratch and mar resistance and lowered coefficient of friction.

Coatings

Silicone films can be applied to such silica-based substrates as glass to form a covalently bonded hydrophobic coating.

Many fabrics can be coated or impregnated with silicone to form a strong, waterproof composite such as silnylon.

Cookware

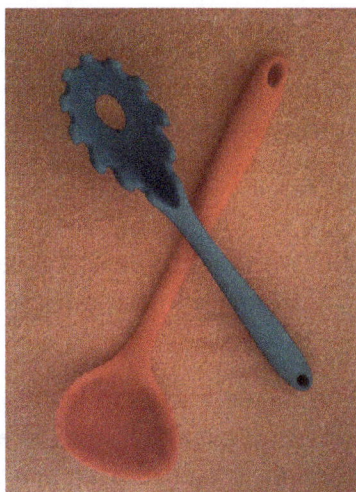

Soup ladle and pasta ladle made of silicone.

A silicone food steamer to be placed inside a pot of boiling water.

Ice cube trays made of silicone.

- As a low-taint, non-toxic material, silicone can be used where contact with food is required. Silicone is becoming an important product in the cookware industry, particularly bakeware and kitchen utensils.

- Silicone is used as an insulator in heat-resistant potholders and similar items; however, it is more conductive of heat than similar less dense fiber-based products. Silicone oven mitts are able to withstand temperatures up to 260 °C (500 °F), allowing reaching into boiling water.

- Molds for chocolate, ice, cookies, muffins and various other foods.

- Non-stick bakeware and reusable mats used on baking sheets.

- Other products such as steamers, egg boilers or poachers, cookware lids, pot holders, trivets, and kitchen mats.

Defoaming

Silicones are used as active compound in defoamers due to their low water solubility and good spreading properties.

Dry Cleaning

Liquid silicone can be used as a dry cleaning solvent, providing an alternative to the traditional chlorine-containing perchloroethylene (perc) solvent. Use of silicones in dry cleaning reduces the environmental impact of a typically high-polluting industry.

Electronics

Electronic components are sometimes encased in silicone to increase stability against mechanical and electrical shock, radiation and vibration, a process called "potting".

Silicones are used where durability and high performance are demanded of components under hard conditions, such as in space (satellite technology). They are selected over polyurethane or epoxy encapsulation when a wide operating temperature range is required (−65 to 315 °C). Silicones also have the advantage of little exothermic heat rise during cure, low toxicity, good electrical properties and high purity.

The use of silicones in electronics is not without problems, however. Silicones are relatively expensive and can be attacked by solvents. Silicone easily migrates as either a liquid or vapor onto other components.

Silicone contamination of electrical switch contacts can lead to failures by causing an increase in contact resistance, often late in the life of the contact, well after any testing is completed. Use of silicone-based spray products in electronic devices during maintenance or repairs can cause later failures.

Firestops

Silicone foam has been used in North American buildings in an attempt to firestop openings within fire-resistance-rated wall and floor assemblies to prevent the spread of flames and smoke from one room to another. When properly installed, silicone-foam firestops can be fabricated for building code compliance. Advantages include flexibility and high dielectric strength. Disadvantages include combustibility (hard to extinguish) and significant smoke development.

Silicone-foam firestops have been the subject of controversy and press attention due to smoke development from pyrolysis of combustible components within the foam, hydrogen gas escape, shrinkage, and cracking. These problems have led to reportable events among licensees (operators of nuclear power plants) of the Nuclear Regulatory Commission (NRC).

Silicone firestops are also used in aircraft.

| Silicone "foamfixer" pump used to apply silicone foam firestop materials. | Self-leveling silicone firestop system used around copper pipe through-penetrations in a two-hour fire-resistance rated concrete floor assembly. | Faulty silicone foam firestop installation in Calgary Sewage Treatment Plant in Canada in the 1980s, attempting to seal the opening above a fire door in a cast concrete fire separation, but improperly set due to wide temperature variations. |

Lubricants

Silicone greases are used for many purposes, such as bicycle chains, airsoft gun parts, and a wide range of other mechanisms. Typically, a dry-set lubricant is delivered with a solvent carrier to penetrate the mechanism. The solvent then evaporates, leaving a clear film that lubricates but does not attract dirt and grit as much as an oil-based or other traditional "wet" lubricant.

Silicone personal lubricants are also available for use in medical procedures or sexual activity.

Medicine

Silicone is used in microfluidics, seals, gaskets, shrouds, and other applications requiring high biocompatibility. Additionally, the gel form is used in bandages and dressings, breast implants, testicle implants, pectoral implants, contact lenses, and a variety of other medical uses.

Scar treatment sheets are often made of medical grade silicone due to its durability and biocompatibility. Polydimethylsiloxane is often used for this purpose, since its specific crosslinking results in a flexible and soft silicone with high durability and tack.

Polydimethylsiloxane (PDMS) has been used as the hydrophobic block of amphiphilic synthetic block copolymers used to form the vesicle membrane of polymersomes.

Moldmaking

Two-part silicone systems are used as rubber molds to cast resins, foams, rubber, and low-temperature alloys. A silicone mold generally requires little or no mold-release or surface preparation, as most materials do not adhere to silicone. For experimental uses, ordinary one-part silicone can be used to make molds or to mold into shapes. If needed, common vegetable cooking oils or petroleum jelly can be used on mating surfaces as a mold-release agent.

Silicone Mold

Cooking molds used as bakeware do not require coating with cooking oil, allowing the baked food to be more easily removed from the mold after cooking.

Ophthalmology

Silicone has many applications like silicone oil used to replace vitreous following vitrectomy, silicone intraocular lenses following cataract extraction, silicone tubes to keep nasolacrimal passage open following dacrycystorhinostomy, canalicular stents for canalicular stenosis, punctal plugs for punctal occlusion in dry eyes, silicone rubber and bands as an external temponade in tractional retinal detachment, and anteriorly located break in rhegmatogenous retinal detachment.

Personal Care

Silicones are ingredients widely used in skin care, color cosmetic and hair care applications. Some silicones, notably the amine functionalized amodimethicones, are excellent conditioners, providing improved compatibility, feel, and softness, and lessening frizz. The phenyltrimethicones, in another silicone family, are used in reflection-enhancing and color-correcting hair products, where they increase shine and glossiness (and possibly effect subtle color changes). Phenyltrimethicones, unlike the conditioning amodimethicones, have refractive indices (typically 1.46) close to that of human hair (1.54). However, if included in the same formulation, amodimethicone and phenyltrimethicone interact and dilute each other, making it difficult to achieve both high shine and excellent conditioning in the same product.

Silicone rubber is commonly used in baby bottle nipples (teats) for its cleanliness, aesthetic appearance, and low extractable content.

Silicones are used in shaving products and personal lubricants.

Plumbing and Building Construction

The strength and reliability of silicone rubber is widely acknowledged in the construction industry. One-part silicone sealants and caulks are in common use to seal gaps, joints and crevices in buildings. One-part silicones cure by absorbing atmospheric moisture, which simplifies installation. In plumbing, silicone grease is typically applied to O-rings in brass taps and valves, preventing lime from sticking to the metal.

Toys and Hobbies

Silly Putty and similar materials are composed of silicones dimethyl siloxane, polydimethylsiloxane, and decamethyl cyclopentasiloxane, with other ingredients. This substance is noted for its unusual characteristics, e.g., that it bounces, but breaks when given a sharp blow; it can also flow like a liquid and will form a puddle given enough time.

Silicone "rubber bands" are a long-lasting popular replacement refill for real rubber bands in the 2013 fad "rubber band loom" toys at two to four times the price (in 2014). Silicone bands also come in bracelet sizes that can be custom embossed with a name or message. Large silicone bands are also sold as utility tie-downs.

Formerol is a silicone rubber (marketed as Sugru) used as an arts-and-crafts material, as its plasticity allows it to be moulded by hand like modeling clay. It hardens at room temperature and it is adhesive to various substances including glass and aluminum.

In making aquariums, manufacturers now commonly use 100% silicone sealant to join glass plates. Glass joints made with silicone sealant can withstand great pressure, making obsolete the original aquarium construction method of angle-iron and putty. This same silicone is used to make hinges in aquarium lids or for minor repairs. However, not all commercial silicones are safe for aquarium manufacture, nor is silicone used for the manufacture of acrylic aquariums as silicones do not have long-term adhesion to plastics.

Sex toys and Lubricants

Silicone is a material of choice for soft sex toys, due to its durability, cleanability, non-degradation by petroleum-based lubricants, and lack of phthalates, chemicals suspected of having carcinogenic and mutagenic effects on the skin and mucous membranes.

Production and Marketing

The global demand for silicones approached US$12.5 billion in 2008, approximately 4% up from the previous year. It continues similar growth in the following years to reach $13.5 billion by 2010. The annual growth is expected to be boosted by broader applications, introduction of novel products and increasing awareness of using more environmentally friendly materials.

The leading global manufacturers of silicone base materials belong to three regional organizations: the European Silicone Center (CES) in Brussels, Belgium; the Environment Health and Safety Council (SEHSC) in Herndon, Virginia, USA; and the Silicone Industry Association of Japan (SIAJ) in Tokyo, Japan. Dow Corning Silicones, Evonik Industries, Momentive Performance Materials, Milliken and Company (SiVance Specialty Silicones), Shin-Etsu Silicones, Wacker Chemie, Bluestar Silicones, JNC Corporation, Wacker Asahikasei Silicone, and Dow Corning Toray represent the collective membership of these organizations. A fourth organization, the Global Silicone Council (GSC) acts as an umbrella structure over the regional organizations. All four are non-profit, having no commercial role; their primary missions are to promote the safety of silicones from a health, safety, and environmental perspective. As the European chemical industry is preparing to implement the Registration, Evaluation and Authorisation of Chemicals (REACH) legislation, CES is leading the formation of a consortium of silicones, silanes, and siloxanes producers and importers to facilitate data and cost sharing.

Safety and Environmental Considerations

No "marked harmful effects on organisms in the environment" have been noted for silicones. Because they are widely used, they are pervasive. They biodegrade readily, a process that is accelerated by a variety of catalysts, including clays.

Around 200 °C in oxygen-containing atmosphere, PDMS releases traces of formaldehyde but less than other common materials such as polyethylene., and by 200 °C (392 °F) Silicones (< 3 µg CH_2O/(g·hr) for a high consistency silicone rubber to 48 µg CH_2O/(g·hr)) were found to be superior to mineral oil and plastics (~400 µg CH_2O/(g·hr)) at about 200 °C (392 °F), by 250 °C (482 °F) copious amounts of formaldehyde have been found to be produced for all silicones (1200 µg CH_2O/(g·hr) to 4600 µg CH_2O/(g·hr)).

Smart Polymer

Smart polymers or stimuli-responsive polymers are high-performance polymers that change according to the environment they are in. Such materials can be sensitive to a number of factors, such as temperature, humidity, pH, the wavelength or intensity of light or an electrical or magnetic field and can respond in various ways, like altering colour or transparency, becoming conductive or permeable to water or changing shape (shape memory polymers). Usually, slight changes in the environment are sufficient to induce large changes in the polymer's properties.

Smart polymers appear in highly specialised applications and everyday products alike. They are used for the production of hydrogels, biodegradable packaging, and to a great extent in biomedical engineering. One example is a polymer that undergoes conformational change in response to pH change, which can be used in drug delivery.

The nonlinear response of smart polymers is what makes them so unique and effective. A significant change in structure and properties can be induced by a very small stimulus. Once that change occurs, there is no further change, meaning a predictable all-or-nothing response occurs, with complete uniformity throughout the polymer. Smart polymers may change conformation, adhesiveness or water retention properties, due to slight changes in pH, ionic strength, temperature or other triggers.

Another factor in the effectiveness of smart polymers lies in the inherent nature of polymers in general. The strength of each molecule's response to changes in stimuli is the composite of changes of individual monomer units which, alone, would be weak. However, these weak responses, compounded hundreds or thousands of times, create a considerable force for driving biological processes.

Stimuli

Several polymer systems respond to temperature, undergoing an lower critical solution temperature phase transition. One of the better-studied such polymers is poly(N-isopropylacryamide), with a transition temperature of approximately 33 °C. Several homologous N-alkyl acrylamides also show LCST behavior, with the transition temperature depending on the length of the hydrophobic side chain. Above their transition temperature, these polymers become insoluble in water. This behavior is believed to be entropy driven.

Classification and Chemistry

Currently, the most prevalent use for smart polymers in biomedicine is for specifically targeted drug delivery. Since the advent of timed-release pharmaceuticals, scientists have been faced with the problem of finding ways to deliver drugs to a particular site in the body without having them first degrade in the highly acidic stomach environment. Prevention of adverse effects to healthy bone and tissue is also an important consideration. Researchers have devised ways to use smart polymers to control the release of drugs until the delivery system has reached the desired target. This release is controlled by either a chemical or physiological trigger.

Linear and matrix smart polymers exist with a variety of properties depending on reactive functional groups and side chains. These groups might be responsive to pH, temperature, ionic strength, electric or magnetic fields, and light. Some polymers are reversibly cross-linked by noncovalent bonds that can break and reform depending on external conditions. Nanotechnology has been fundamental in the development of certain nanoparticle polymers such as dendrimers and fullerenes, that have been applied for drug delivery. Traditional drug encapsulation has been done using lactic acid polymers. More recent developments have seen the formation of lattice-like matrices that hold the drug of interest integrated or entrapped between the polymer strands.

Smart polymer matrices release drugs by a chemical or physiological structure-altering reaction, often a hydrolysis reaction resulting in cleavage of bonds and release of drug as the matrix breaks down into biodegradable components. The use of natural polymers has given way to artificially synthesized polymers such as polyanhydrides, polyesters, polyacrylic acids, poly(methyl methacrylates), and polyurethanes. Hydrophilic, amorphous, low-molecular-weight polymers containing heteroatoms (i.e., atoms other than carbon) have been found to degrade fastest. Scientists control the rate of drug delivery by varying these properties thus adjusting the rate of degradation.

A graft-and-block copolymer is two different polymers grafted together. A number of patents already exist for different combinations of polymers with different reactive groups. The product exhibits properties of both individual components which adds a new dimension to a smart polymer structure, and may be useful for certain applications. Cross-linking hydrophobic and hydrophilic polymers results in formation of micelle-like structures that can protectively assist drug delivery through aqueous medium until conditions at the target location cause simultaneous breakdown of both polymers.

A graft-and-block approach might be useful for solving problems encountered by the use of a common bioadhesive polymer, polyacrylic acid (PAAc). PAAc adheres to mucosal surfaces but will swell and degrade rapidly at pH 7.4, resulting in rapid release of drugs entrapped in its matrix. A combination of PAAc with another polymer that is less sensitive to changes at neutral pH might increase the residence time and slow the release of the drug, thus improving bioavailability and effectiveness.

Hydrogels are polymer networks that do not dissolve in water but swell or collapse in changing aqueous environments. They are useful in biotechnology for phase separation because they are reusable or recyclable. New ways to control the flow, or catch and release of target compounds, in hydrogels, are being investigated. Highly specialized hydrogels have been developed for the delivery and release of drugs into specific tissues. Hydrogels made from PAAc are especially common because of their bioadhesive properties and tremendous absorbency.

Enzyme immobilization in hydrogels is a fairly well-established process. Reversibly cross-linked polymer networks and hydrogels can be similarly applied to a biological system where the response and release of a drug is triggered by the target molecule itself. Alternatively, the response might be turned on or off by the product of an enzyme reaction. This is often done by incorporating an enzyme, receptor or antibody, that binds to the molecule of interest, into the hydrogel. Once bound, a chemical reaction takes place that triggers a reaction from the hydrogel. The trigger can be oxygen, sensed using oxidoreductase enzymes, or a pH-sensing response. An example of the latter is combined entrapment of glucose oxidase and insulin in a pH-responsive hydrogel. In the presence of glucose, the formation of gluconic acid by the enzyme triggers release of insulin from the hydrogel.

Two criteria for this technology to work effectively are enzyme stability and rapid kinetics (quick response to the trigger and recovery after removal of the trigger). Several strategies have been tested in type 1 diabetes research, involving the use of similar types of smart polymers that can detect changes in blood glucose levels and trigger production or release of insulin. Likewise, there are many possible applications of similar hydrogels as drug delivery agents for other conditions and diseases.

Other Applications

Smart polymers are not just for drug delivery. Their properties make them especially suited for bioseparations. The time and costs involved in purifying proteins might be reduced significantly by using smart polymers that undergo rapid reversible changes in response to a change in medium properties. Conjugated systems have been used for many years in physical and affinity separations and immunoassays. Microscopic changes in the polymer structure are manifested as precipitate formation, which may be used to aid separation of trapped proteins from solution.

These systems work when a protein or other molecule that is to be separated from a mix, forms a bioconjugate with the polymer, and precipitates with the polymer when its environment undergoes a change. The precipitate is removed from the media, thus separating the desired component of the conjugate from the rest of the mixture. Removal of this component from the conjugate depends on recovery of the polymer and a return to its original state, thus hydrogels are very useful for such processes.

Another approach to controlling biological reactions using smart polymers is to prepare recombinant proteins with built-in polymer binding sites close to ligand or cell binding sites. This technique has been used to control ligand and cell binding activity, based on a variety of triggers including temperature and light.

Future Applications

It has been suggested that polymers might be developed that can learn and self-correct behavior over time. Although this might be a far distant possibility, there are other more feasible applications that appear to be coming in the near future. One of these is the idea of smart toilettes that analyze urine and help identify health problems. In environmental biotechnology, smart irrigation systems have been also been proposed. It would be incredibly useful to have a system that turns on and off, and controls fertilizer concentrations, based on soil moisture, pH and nutrient levels. Many creative approaches to targeted drug delivery systems that self-regulate based on their unique cellular surroundings, are also under investigation.

There are obvious possible problems associated with the use of smart polymers in biomedicine. The most worrisome is the possibility of toxicity or incompatibility of artificial substances in the body, including degradation products and byproducts. However, smart polymers have enormous potential in biotechnology and biomedical applications if these obstacles can be overcome.

Polystyrene

Polystyrene (PS) is a synthetic aromatic polymer made from the monomer styrene. Polystyrene can be solid or foamed. General-purpose polystyrene is clear, hard, and rather brittle. It is an inexpensive resin per unit weight. It is a rather poor barrier to oxygen and water vapor and has a relatively low melting point. Polystyrene is one of the most widely used plastics, the scale of its production being several billion kilograms per year. Polystyrene can be naturally transparent, but can be colored with colorants. Uses include protective packaging (such as packing peanuts and CD and DVD cases), containers (such as "clamshells"), lids, bottles, trays, tumblers, and disposable cutlery.

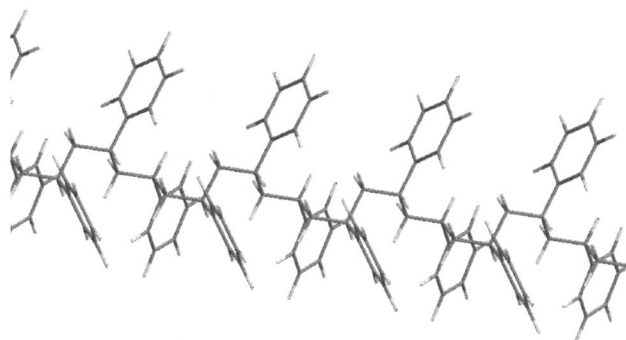

Stick model of polystyrene

Expanded polystyrene packaging

A polystyrene yogurt container

Bottom of a vacuum-formed cup; note how fine details such as the glass and fork food contact materials symbol and the resin identification code symbol are easily molded

As a thermoplastic polymer, polystyrene is in a solid (glassy) state at room temperature but flows if heated above about 100 °C, its glass transition temperature. It becomes rigid again when cooled. This temperature behavior is exploited for extrusion (as in Styrofoam) and also for molding and vacuum forming, since it can be cast into molds with fine detail.

Polystyrene is very slow to biodegrade and is therefore a focus of controversy among environmentalists. It is increasingly abundant as a form of litter in the outdoor environment, particularly along shores and waterways, especially in its foam form, and also in increasing quantities in the Pacific Ocean.

History

Polystyrene was discovered in 1839 by Eduard Simon, an apothecary from Berlin. From storax, the resin of the American sweetgum tree *Liquidambar styraciflua*, he distilled an oily substance, a monomer that he named styrol. Several days later, Simon found that the styrol had thickened into a jelly he dubbed styrol oxide ("Styroloxyd") because he presumed an oxidation. By 1845 Jamaican-born chemist John Buddle Blyth and German chemist August Wilhelm von Hofmann showed that the same transformation of styrol took place in the absence of oxygen. They called the product "metastyrol"; analysis showed that it was chemically identical to Simon's Styroloxyd. In 1866 Marcelin Berthelot correctly identified the formation of metastyrol/Styroloxyd from styrol as a polymerization process. About 80 years later it was realized that heating of styrol starts a chain reaction that produces macromolecules, following the thesis of German organic chemist Hermann Staudinger (1881–1965). This eventually led to the substance receiving its present name, polystyrene.

The company I. G. Farben began manufacturing polystyrene in Ludwigshafen, about 1931, hoping it would be a suitable replacement for die-cast zinc in many applications. Success was achieved when they developed a reactor vessel that extruded polystyrene through a heated tube and cutter, producing polystyrene in pellet form.

In 1941, Dow Chemical invented a Styrofoam process.

Before 1949, the chemical engineer Fritz Stastny (1908–1985) developed pre-expanded PS beads by incorporating aliphatic hydrocarbons, such as pentane. These beads are the raw material for moulding parts or extruding sheets. BASF and Stastny applied for a patent that was issued in 1949. The moulding process was demonstrated at the Kunststoff Messe 1952 in Düsseldorf. Products were named Styropor.

The crystal structure of isotactic polystyrene was reported by Giulio Natta.

In 1954, the Koppers Company in Pittsburgh, Pennsylvania, developed expanded polystyrene (EPS) foam under the trade name Dylite.

In 1960, Dart Container, the largest manufacturer of foam cups, shipped their first order.

In 1988, the first U.S. ban of general polystyrene foam was enacted in Berkeley, California.

Structure

Polystyrene is flammable.

In chemical terms, polystyrene is a long chain hydrocarbon wherein alternating carbon centers are attached to phenyl groups (the name given to the aromatic ring benzene). Polystyrene's chemical formula is $(C_8H_8)_n$; it contains the chemical elements carbon and hydrogen.

The material's properties are determined by short-range van der Waals attractions between polymers chains. Since the molecules are long hydrocarbon chains that consist of thousands of atoms, the total attractive force between the molecules is large. When heated (or deformed at a rapid rate, due to a combination of viscoelastic and thermal insulation properties), the chains are able to take on a higher degree of conformation and slide past each other. This intermolecular weakness (versus the high *intramolecular* strength due to the hydrocarbon backbone) confers flexibility and elasticity. The ability of the system to be readily deformed above its glass transition temperature allows polystyrene (and thermoplastic polymers in general) to be readily softened and molded upon heating.

Extruded polystyrene is about as strong as an unalloyed aluminium, but much more flexible and much lighter (1.05 g/cm^3 vs. 2.70 g/cm^3 for aluminium).

Polymerization

Polystyrene results when styrene monomers interconnect. In the polymerization, the carbon-carbon pi bond (in the vinyl group) is broken and a new carbon-carbon single (sigma) bond is formed, attaching another styrene monomer to the chain. The newly formed sigma bond is much stronger than the pi bond that was broken, thus it is very difficult to depolymerize polystyrene. About a few thousand monomers typically comprise a chain of polystyrene, giving a molecular weight of 100,000–400,000.

A 3-D model would show that each of the chiral backbone carbons lies at the center of a tetrahedron, with its 4 bonds pointing toward the vertices. Consider that the -C-C- bonds are rotated so that the backbone chain lies entirely in the plane of the diagram. From this flat schematic, it is not evident which of the phenyl (benzene) groups are angled outward from the plane of the diagram, and which ones are inward. The isomer where all of the phenyl groups are on the same side is called *isotactic* polystyrene, which is not produced commercially.

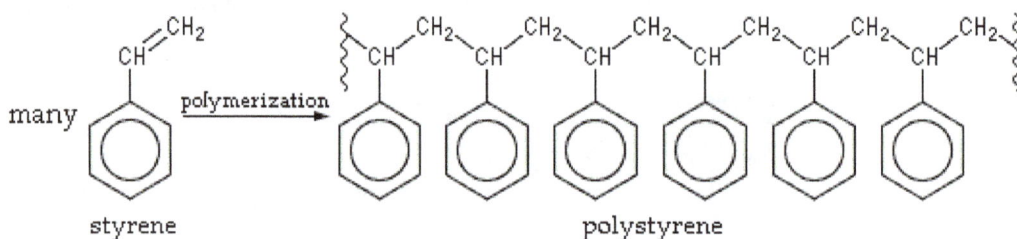

styrene   polystyrene

Atactic Polystyrene

The only commercially important form of polystyrene is *atactic*, in which the phenyl groups are randomly distributed on both sides of the polymer chain. This random positioning prevents the chains from aligning with sufficient regularity to achieve any crystallinity. The plastic has a glass transition temperature T_g of ~90 °C. Polymerization is initiated with free radicals.

Syndiotactic Polystyrene

Ziegler-Natta polymerization can produce an ordered *syndiotactic* polystyrene with the phenyl groups positioned on alternating sides of the hydrocarbon backbone. This form is highly crystalline with a T_m of 270 °C (518 °F). Syndiotactic polystyrene resin is currently produced under the trade name XAREC by Idemitsu corporation. Syndiotactic polystyrene is prepared by combining a metallocene catalyst with a styrene monomer to generate a polystyrene chain with a syndiotactic structure.

Degradation

Polystyrene is chemically very inert, being resistant to acids and bases but is easily dissolved by many chlorinated solvents, and many aromatic hydrocarbon solvents. Because of its resilience and inertness, it is used for fabricating many objects of commerce. It is attacked by many organic solvents, which dissolve the polymer. Foamed polystyrene is used for packaging chemicals.

Like all organic compounds, polystyrene burns to give carbon dioxide and water vapor. Polystyrene, being an aromatic hydrocarbon, typically combusts incompletely as indicated by the sooty flame.

Biodegradation

Polystyrene is generally non-biodegradable. There are a couple of exceptions:

To quote:

Methanogenic consortia have been shown to degrade styrene as sole carbon source (Grbić-Galić et al. 1990). In this case styrene degraded to a range of organic intermediates and carbon dioxide. Taking the carbon dioxide figures as a representation of the amount of styrene that had completely degraded to gas as is of interest here, rates of styrene degradation ranged from 0.14 to 0.4 a^{-1}. This is an order of magnitude faster than the most rapid rate of polystyrene degradation identified (Kaplan et al. 1979, Sielicki et al. 1978). It is consistent with the T2GGM polystyrene degradation model (Quintessa and Geofirma 2011b), which considers the rate-limiting step for polystyrene degradation to be the breakup of polystyrene, rather than the degradation of styrene.

Mealworms have been shown to be able to eat polystyrene and degrade it within their larval gut.

Pseudomonas putida is capable of converting styrene oil into the biodegradable plastic PHA. This may someday be of use in the effective disposing of polystyrene foam.

Forms Produced

Properties	
Density of EPS	16–640 kg/m³
Young's modulus (E)	3000–3600 MPa
Tensile strength (s_t)	46–60 MPa
Elongation at break	3–4%
Notch test	2–5 kJ/m²
Glass transition temperature	100 °C
Vicat B	90 °C

Linear expansion coefficient (a)	8×10^{-5} /K
Specific heat (c)	1.3 kJ/(kg·K)
Water absorption (ASTM)	0.03–0.1
Decomposition	X years, still decaying

Polystyrene is commonly injection molded, vacuum formed, or extruded, while expanded polystyrene is either extruded or molded in a special process. Polystyrene copolymers are also produced; these contain one or more other monomers in addition to styrene. In recent years the expanded polystyrene composites with cellulose and starch have also been produced. Polystyrene is used in some polymer-bonded explosives (PBX).

Sheet or Molded Polystyrene

Polystyrene (PS) is used for producing disposable plastic cutlery and dinnerware, CD "jewel" cases, smoke detector housings, license plate frames, plastic model assembly kits, and many other objects where a rigid, economical plastic is desired. Production methods include thermoforming (vacuum forming) and injection molding.

CD case made from general purpose polystyrene (GPPS) and high impact polystyrene (HIPS)

Disposable polystyrene razor

Polystyrene Petri dishes and other laboratory containers such as test tubes and microplates play an important role in biomedical research and science. For these uses, articles are almost always made by injection molding, and often sterilized post-molding, either by irradiation or by treatment with ethylene oxide. Post-mold surface modification, usually with oxygen-rich plasmas, is often done to introduce polar groups. Much of modern biomedical research relies on the use of such products; they therefore play a critical role in pharmaceutical research.

Foams

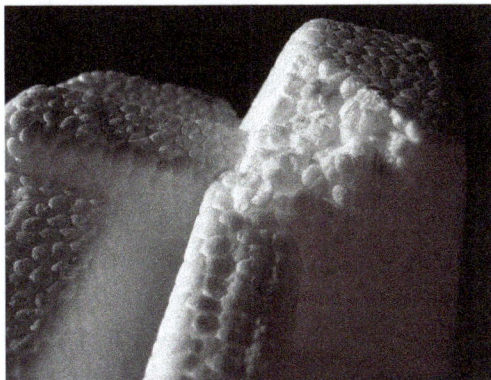

Close up of expanded polystyrene packaging

Polystyrene foams are good thermal insulators and are therefore often used as building insulation materials, such as in insulating concrete forms and structural insulated panel building systems. Grey polystyrene foam, incorporating graphite has superior insulation properties. They are also used for non-weight-bearing architectural structures (such as ornamental pillars). PS foams also exhibit good damping properties, therefore it is used widely in packaging. The trademark Styrofoam by Dow Chemical Company is informally used (mainly US & Canada) for all foamed polystyrene products, although strictly it should only be used for 'extruded closed-cell' polystyrene foams made by Dow Chemicals.

Expanded Polystyrene (EPS)

Expanded polystyrene (EPS) is a rigid and tough, closed-cell foam. It is usually white and made of pre-expanded polystyrene beads. EPS is used for many applications e.g. trays, plates, bowls and fish boxes. Other uses include molded sheets for building insulation and packing material ("peanuts") for cushioning fragile items inside boxes. Sheets are commonly packaged as rigid panels (size 4 by 8 or 2 by 8 feet in the United States), which are also known as "bead-board".

Due to its technical properties such as low weight, rigidity, and formability, EPS can be used in a wide range of different applications. Its market value is likely to rise to more than US$15 billion by 2020.

Thermal conductivity is measured according to EN 12667. Typical values range from 0.032 to 0.038 W/(m·K) depending on the density of the EPS board. The value of 0.038 W/(m·K) was obtained at 15 kg/m^3 while the value of 0.032 W/(m·K) was obtained at 40 kg/m^3 according to the data sheet of K-710 from StyroChem Finland. Adding fillers (graphites, aluminium, or carbons) has recently allowed the thermal conductivity of EPS to reach around 0.030–0.034 (as low as 0.029) and as such has a grey/black color which distinguishes it from standard EPS. Several EPS producers have produced a variety of these increased thermal resistance EPS usage for this product in the UK & EU.

Water vapor diffusion resistance (μ) of EPS is around 30–70.

ICC-ES (International Code Council Evaluation Service) requires EPS boards used in building construction meet ASTM C578 requirements. One of these requirements is that the oxygen index of EPS as measured by ASTM D2863 be greater than 24 volume %. Typical EPS has an oxygen index of around 18 volume %; thus, a flame retardant is added to styrene or polystyrene during the formation of EPS.

The boards containing a flame retardant when tested in a tunnel using test method UL 723 or ASTM E84 will have a flame spread index of less than 25 and a smoke-developed index of less than 450. ICC-ES requires the use of a 15-minute thermal barrier when EPS boards are used inside of a building.

According to EPS-IA ICF organization, the typical density of EPS used for insulated concrete forms is 1.35 to 1.80 pcf. This is either Type II or Type IX EPS according to ASTM C578. EPS blocks or boards used in building construction are commonly cut using hot wires.

Extruded Polystyrene Foam

Extruded polystyrene foam (XPS) consists of closed cells, offers improved surface roughness and higher stiffness and reduced thermal conductivity. The density range is about 28–45 kg/m³.

Extruded polystyrene material is also used in crafts and model building, in particular architectural models. Because of the extrusion manufacturing process, XPS does not require facers to maintain its thermal or physical property performance. Thus, it makes a more uniform substitute for corrugated cardboard. Thermal conductivity varies between 0.029 and 0.039 W/(m·K) depending on bearing strength/density and the average value is ~0.035 W/(m·K).

Water vapour diffusion resistance (μ) of XPS is around 80–250 and so makes it more suitable to wetter environments than EPS.

Water Absorption of Polystyrene Foams

Although it is a closed-cell foam, both expanded and extruded polystyrene are not entirely waterproof or vaporproof. In expanded polystyrene there are interstitial gaps between the expanded closed-cell pellets that form an open network of channels between the bonded pellets, and this network of gaps can become filled with liquid water. If the water freezes into ice, it expands and can cause polystyrene pellets to break off from the foam. Extruded polystyrene is also permeable by water molecules and can not be considered a vapor barrier.

Waterlogging commonly occurs over a long period of time in polystyrene foams that are constantly exposed to high humidity or are continuously immersed in water, such as in hot tub covers, in floating docks, as supplemental flotation under boat seats, and for below-grade exterior building insulation constantly exposed to groundwater. Typically an exterior vapor barrier such as impermeable plastic sheeting or a sprayed-on coating is necessary to prevent saturation.

Copolymers

Pure polystyrene is brittle, but hard enough that a fairly high-performance product can be made by giving it some of the properties of a stretchier material, such as polybutadiene rubber. The two such materials can never normally be mixed because of the small mixing entropy of polymers, but if polybutadiene is added during polymerization it can become chemically bonded to the polystyrene, forming a graft copolymer, which helps to incorporate normal polybutadiene into the final mix, resulting in high-impact polystyrene or HIPS, often called "high-impact plastic" in advertisements. One commercial name for HIPS is Bextrene. Common applications of HIPS include toys and product casings. HIPS is usually injection molded in production. Autoclaving polystyrene can compress and harden the material.

Several other copolymers are also used with styrene. Acrylonitrile butadiene styrene or ABS plastic is similar to HIPS: a copolymer of **a**crylonitrile and **s**tyrene, toughened with poly**b**utadiene. Most electronics cases are made of this form of polystyrene, as are many sewer pipes. SAN is a copolymer of styrene with acrylonitrile, and SMA one with maleic anhydride. Styrene can be copolymerized with other monomers; for example, divinylbenzene can be used for cross-linking the polystyrene chains to give the polymer used in Solid phase peptide synthesis.

Oriented Polystyrene

Oriented polystyrene (OPS) is produced by stretching extruded PS film, improving visibility through the material by reducing haziness and increasing stiffness. This is often used in packaging where the manufacturer would like the consumer to see the enclosed product. Some benefits to OPS are that it is less expensive to produce than other clear plastics such as PP, PET, and HIPS, and it is less hazy than HIPS or PP. The main disadvantage to OPS is that it is brittle, and will crack or tear easily.

Environmental Issues

Production

Polystyrene foams are produced using blowing agents that form bubbles and expand the foam. In expanded polystyrene, these are usually hydrocarbons such as pentane, which may pose a flammability hazard in manufacturing or storage of newly manufactured material, but have relatively mild environmental impact. Extruded polystyrene is usually made with hydrofluorocarbons (HFC-134a), which have global warming potentials of approximately 1000–1300 times that of carbon dioxide.

Non-biodegradable

Discarded polystyrene does not biodegrade for hundreds of years and is resistant to photolysis.

Litter

Animals do not recognize polystyrene foam as an artificial material and may even mistake it for food. Polystyrene foam blows in the wind and floats on water, due to its specific gravity. It can have serious effects on the health of birds or marine animals that swallow significant quantities.

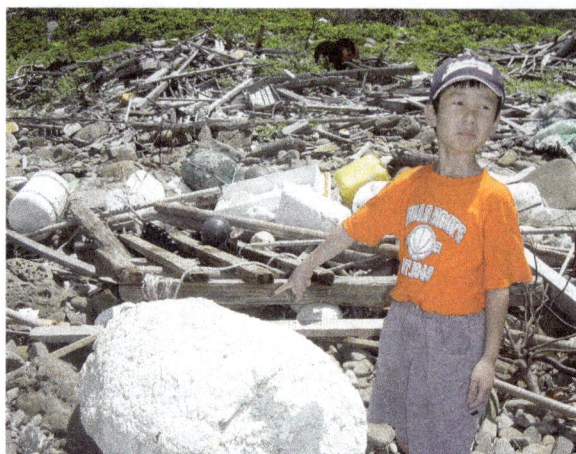
Coastal debris including polystyrene in Japan

Reducing

Restricting the use of foamed polystyrene takeout food packaging is a priority of many solid waste environmental organizations. Efforts have been made to find alternatives to polystyrene, especially foam in restaurant settings. The original impetus was to eliminate chlorofluorocarbons (CFC), which was a former component of foam.

United States

In 1987, Berkeley, California banned CFC food containers. The following year, Suffolk County, New York became the first U.S. locality to ban polystyrene. However, legal challenges by the Society of the Plastics Industry kept it from going into effect until at last it was delayed when the Republican and Conservative parties became a majority of the county legislature. In the meantime, Berkeley became the first city to ban all foam food containers. As of 2006, about one hundred localities in the United States including Portland, Oregon and San Francisco currently have some sort of ban on polystyrene foam in restaurants. For instance, in 2007 Oakland, California required restaurants to switch to disposable food containers that will biodegrade if added to food compost. In 2013, San Jose became reportedly the largest city in the country to ban polystyrene foam food containers. Some communities have implemented wide polystyrene bans, such as Freeport, Maine, which did so in 1990.

The U.S. Green Restaurant Association does not allow polystyrene foam to be used as part of their certification standard. Several green leaders, from the Dutch Ministry of the Environment to Starbucks' Green Team, advise that individuals reduce their environmental impact by using reusable coffee cups.

Outside the United States

China banned expanded polystyrene takeout/takeaway containers and tableware around 1999. However, compliance has been a problem and, in 2013, the Chinese plastics industry is actively lobbying to get the ban repealed.

India and Taiwan also banned polystyrene foam food service ware prior to 2007.

Recycling

The resin identification code symbol for polystyrene

In general, polystyrene is not accepted in curbside collection recycling programs, and is not separated and recycled where it is accepted. In Germany, polystyrene is collected, as a consequence of the packaging law (Verpackungsverordnung) that requires manufacturers to take responsibility for recycling or disposing of any packaging material they sell.

Most polystyrene products are currently not recycled due to the lack of incentive to invest in the compactors and logistical systems required. Due to the low density of polystyrene foam, it is not economical to collect. However, if the waste material goes through an initial compaction process, the material changes density from typically 30 kg/m³ to 330 kg/m³ and becomes a recyclable commodity of high value for producers of recycled plastic pellets. Expanded polystyrene scrap can be easily added to products such as EPS insulation sheets and other EPS materials for construction applications; many manufacturers cannot obtain sufficient scrap because of collection issues. When it is not used to make more EPS, foam scrap can be turned into products such as clothes hangers, park benches, flower pots, toys, rulers, stapler bodies, seedling containers, picture frames, and architectural molding from recycled PS. Currently, around 100 tonnes of EPS are recycled every month in the UK.

Recycled EPS is also used in many metal casting operations. Rastra is made from EPS that is combined with cement to be used as an insulating amendment in the making of concrete foundations and walls. American manufacturers have produced insulating concrete forms made with approximately 80% recycled EPS since 1993.

Incineration

If polystyrene is properly incinerated at high temperatures (up to 1000 °C) and with plenty of air (14 m³/kg), the chemicals generated are water, carbon dioxide, and possibly small amounts of residual halogen-compounds from flame-retardants. If only incomplete incineration is done, there will also be leftover carbon soot and a complex mixture of volatile compounds. According to the American Chemistry Council, when polystyrene is incinerated in modern facilities, the final volume is 1% of the starting volume; most of the polystyrene is converted into carbon dioxide, water vapor, and heat. Because of the amount of heat released, it is sometimes used as a power source for steam or electricity generation.

When polystyrene was burned at temperatures of 800–900 °C (the typical range of a modern incinerator), the products of combustion consisted of "a complex mixture of polycyclic aromatic hydrocarbons (PAHs) from alkyl benzenes to benzoperylene. Over 90 different compounds were identified in combustion effluents from polystyrene."

Safety

Health

According to a plastic food service products website of the American Chemistry Council:

Based on scientific tests over five decades, government safety agencies have determined that polystyrene may be safe for use in foodservice products. For example, polystyrene comes close to meeting the standards of the U.S. Food and Drug Administration and the European Commission/European Food Safety Authority for use in packaging to store and serve food. The Hong Kong Food and Environmental Hygiene Department recently reviewed the safety of serving various foods in polystyrene foodservice products and reached the same conclusion as the U.S. FDA.

From 1999 to 2002, a comprehensive review of the potential health risks associated with exposure to styrene was conducted by a 12-member international expert panel selected by the Harvard Center for Risk Assessment. The scientists had expertise in toxicology, epidemiology, medicine, risk analysis, pharmacokinetics, and exposure assessment.

The Harvard study reported that styrene is naturally present in trace quantities in foods such as strawberries, beef, and spices, and is naturally produced in the processing of foods such as wine and cheese. The study also reviewed all the published data on the quantity of styrene contributing to the diet due to migration of food packaging and disposable food contact articles, and concluded there is cause for limited concern for the general public from exposure to styrene from foods or styrenic materials used in food-contact applications, such as polystyrene packaging and food service containers, especially after microwaving.

Polystyrene is commonly used in containers for food and drinks. The styrene monomer (from which polystyrene is made) is a cancer suspect agent. Styrene is "generally found in such low levels in consumer products that risks aren't substantial". Polystyrene which is used for food contact may not contain more than 1% (0.5% for fatty foods) of styrene by weight. Styrene oligomers in polystyrene containers used for food packaging have been found to migrate into the food. Another Japanese study conducted on wild-type and AhR-null mice found that the styrene trimer, which the authors detected in cooked polystyrene container-packed instant foods, may increase thyroid hormone levels.

Whether polystyrene can be microwaved with food is controversial. Some containers may be safely used in a microwave, but only if labelled as such. Some sources suggest that foods containing carotene (Vitamin A) or cooking oils must be avoided.

Because of the pervasive use of polystyrene, these serious health related issues remain topical.

Fire Hazards

Like other organic compounds, polystyrene is flammable. Polystyrene is classified according to DIN4102 as a "B3" product, meaning highly flammable or "Easily Ignited." As a consequence, although it is an efficient insulator at low temperatures, its use is prohibited in any exposed installations in building construction if the material is not flame-retardant. It must be concealed behind drywall, sheet metal, or concrete. Foamed polystyrene plastic materials have been accidentally ignited and caused huge fires and losses, for example at the Düsseldorf International Airport and the Channel tunnel (where polystyrene was inside a railcar that caught fire).

Polymer Architecture

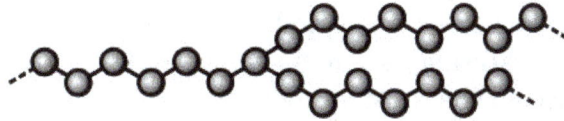

Branch point in a polymer

Polymer architecture in polymer science relates to the way branching leads to a deviation from a strictly linear polymer chain. Branching may occur randomly or reactions may be designed so that specific architectures are targeted. It is an important microstructural feature. A polymer's architecture affects many of its physical properties including solution viscosity, melt viscosity, solubility in various solvents, glass transition temperature and the size of individual polymer coils in solution.

Different Polymer Architectures

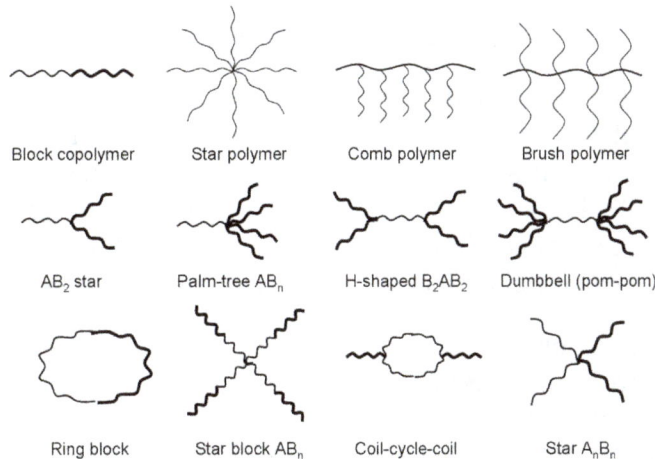

Block copolymer	Star polymer	Comb polymer	Brush polymer
AB$_2$ star	Palm-tree AB$_n$	H-shaped B$_2$AB$_2$	Dumbbell (pom-pom)
Ring block	Star block AB$_n$	Coil-cycle-coil	Star A$_n$B$_n$

Various polymer architectures.

Random Branching

Branches can form when the growing end of a polymer molecule reaches either (a) back around onto itself or (b) onto another polymer chain, both of which, via abstraction of a hydrogen, can create a mid-chain growth site.

Branching can be quantified by the branching index.

Cross Linked Polymer

An effect related to branching is chemical crosslinking - the formation of covalent bonds between chains. Crosslinking tends to increase T_g and increase strength and toughness. Among other applications, this process is used to strengthen rubbers in a process known as vulcanization, which is based on crosslinking by sulfur. Car tires, for example, are highly crosslinked in order to reduce the leaking of air out of the tire and to toughen their durability. Eraser rubber, on the other hand, is not crosslinked to allow flaking of the rubber and prevent damage to the paper. Polymerization of pure sulfur at higher temperatures also explains why sulfur becomes more viscous with elevated temperatures in its molten state.

A polymer molecule with a high degree of crosslinking is referred to as a polymer network. A sufficiently high crosslink to chain ratio may lead to the formation of a so-called infinite network or gel, in which each chain is connected to at least one other.

Complex Architectures

With the continual development of Living polymerization, the synthesis of polymers with specific architectures becomes more and more facile. Architectures such as star polymers, comb polymers, brush polymers, dendronized polymers, dendrimers and Ring polymers are possible. Complex architecture polymers can be synthesized either with the use of specially tailored starting compounds or by first synthesising linear chains which undergo further reactions to become connected together. Knotted polymers consist of multiple intramolecular cyclization units within a single polymer chain.

Effect of Architecture on Physical Properties

In general, the higher degree of branching, the more compact a polymer chain is. Branching also affects chain entanglement, the ability of chains to slide past one another, in turn affecting the bulk physical properties. Long chain branches may increase polymer strength, toughness, and the glass transition temperature (T_g) due to an increase in the number of entanglements per chain. A random and short chain length between branches, on the other hand, may reduce polymer strength due to disruption of the chains' ability to interact with each other or crystallize.

An example of the effect of branching on physical properties can be found in polyethylene. High-density polyethylene (HDPE) has a very low degree of branching, is relatively stiff, and is used in applications such as bullet-proof vests. Low-density polyethylene (LDPE), on the other hand, has significant numbers of both long and short branches, is relatively flexible, and is used in applications such as plastic films.

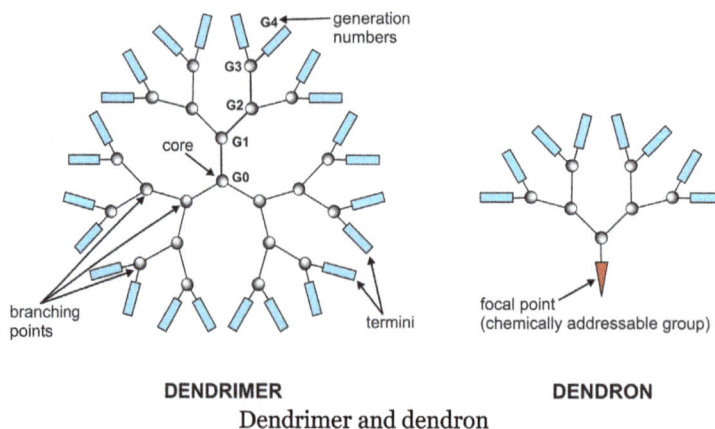

Dendrimer and dendron

Dendrimers are a special case of branched polymer where every monomer unit is also a branch point. This tends to reduce intermolecular chain entanglement and crystallization. A related architecture, the dendritic polymer, are not perfectly branched but share similar properties to dendrimers due to their high degree of branching.

The degree of branching that occurs during polymerisation can be influenced by the functionality of the monomers that are used. For example, in a free radical polymerisation of styrene, addition of divinylbenzene, which has a functionality of 2, will result in the formation of branched polymer.

Polymer Science

Polymer science or macromolecular science is a subfield of materials science concerned with polymers, primarily synthetic polymers such as plastics and elastomers. The field of polymer science includes researchers in multiple disciplines including chemistry, physics, and engineering.

Subdisciplines

This science comprises three main sub-disciplines:

- Polymer chemistry or macromolecular chemistry is concerned with the chemical synthesis and chemical properties of polymers.

- Polymer physics is concerned with the bulk properties of polymer materials and engineering applications.

- Polymer characterization is concerned with the analysis of chemical structure and morphology and the determination of physical properties in relation to compositional and structural parameters.

History of Polymer Science

The earliest known work with polymers was the rubber industry in pre-Columbian Mexico. The mesoamericans knew how to combine latex of the rubber tree with the juice of the morning glory plant in different proportions to get rubber with different properties for different products, such as bouncing balls, sandals, and rubber bands.

The first modern example of polymer science is Henri Braconnot's work in the 1830s. Braconnot, along with Christian Schönbein and others, developed derivatives of the natural polymer cellulose, producing new, semi-synthetic materials, such as celluloid and cellulose acetate. The term "polymer" was coined in 1833 by Jöns Jakob Berzelius, though Berzelius did little that would be considered polymer science in the modern sense. In the 1840s, Friedrich Ludersdorf and Nathaniel Hayward independently discovered that adding sulfur to raw natural rubber (vulcanizing natural rubber with sulfur and heat. Thomas Hancock had received a patent for the same process in the UK the year before. This process strengthened natural rubber and prevented it from melting with heat without losing flexibility. This made practical products such as waterproofed articles possible. It also facilitated practical manufacture of such rubberized materials. Vulcanized rubber represents the first commercially successful product of polymer research. In 1884 Hilaire de Chardonnet started the first artificial fiber plant based on regenerated cellulose, or viscose rayon, as a substitute for silk, but it was very flammable. In 1907 Leo Baekeland invented the first synthetic polymer, a thermosetting phenol–formaldehyde resin called Bakelite.

Despite significant advances in polymer synthesis, the molecular nature of polymers was not understood until the work of Hermann Staudinger in 1922. Prior to Staudinger's work, polymers were understood in terms of the association theory or aggregate theory, which originated with Thomas Graham in 1861. Graham proposed that cellulose and other polymers were colloids, aggregates of molecules having small molecular mass connected by an unknown intermolecular force. Hermann Staudinger was the first to propose that polymers consisted of long chains of atoms held together by covalent bonds. It took over a decade for Staudinger's work to gain wide acceptance in the scientific community, work for which he was awarded the Nobel Prize in 1953.

The World War II era marked the emergence of a strong commercial polymer industry. The limited or restricted supply of natural materials such as silk and rubber necessitated the increased production of synthetic substitutes, such as nylon and synthetic rubber. In the intervening years, the development of advanced polymers such as Kevlar and Teflon have continued to fuel a strong and growing polymer industry.

The growth in industrial applications was mirrored by the establishment of strong academic programs and research institute. In 1946, Herman Mark established the Polymer Research Institute at Brooklyn Polytechnic, the first research facility in the United States dedicated to polymer research. Mark is also recognized as a pioneer in establishing curriculum and pedagogy for the field of polymer science. In 1950, the POLY division of the American Chemical Society was formed, and has since grown to the second-largest division in this association with nearly 8,000 members. Fred W. Billmeyer, Jr., a Professor of Analytical Chemistry had once said that "although the scarcity of education in polymer science is slowly diminishing but it is still evident in many areas. What is most unfortunate is that it appears to exist, not because of a lack of awareness but, rather, a lack of interest."

Nobel Prizes Related to Polymer Science

2005 (Chemistry) Robert Grubbs, Richard Schrock, Yves Chauvin for olefin metathesis.

2002 (Chemistry) John Bennett Fenn, Koichi Tanaka, and Kurt Wüthrich for the development of methods for identification and structure analyses of biological macromolecules.

2000 (Chemistry) Alan G. MacDiarmid, Alan J. Heeger, and Hideki Shirakawa for work on conductive polymers, contributing to the advent of molecular electronics.

1991 (Physics) Pierre-Gilles de Gennes for developing a generalized theory of phase transitions with particular applications to describing ordering and phase transitions in polymers.

1974 (Chemistry) Paul J. Flory for contributions to theoretical polymer chemistry.

1963 (Chemistry) Giulio Natta and Karl Ziegler for contributions in polymer synthesis. (Ziegler-Natta catalysis).

1953 (Chemistry) Hermann Staudinger for contributions to the understanding of macromolecular chemistry.

Polymeric Surface

Polymeric materials have widespread application due to their versatile characteristics, cost-effectiveness, and highly tailored production. The science of polymer synthesis allows for excellent control over the properties of a bulk polymer sample. However, surface interactions of polymer substrates are an essential area of study in biotechnology, nanotechnology, and in all forms of coating applications. In these cases, the surface characteristics of the polymer and material, and the resulting forces between them largely determine its utility and reliability. In biomedical applications for example, the bodily response to foreign material, and thus biocompatibility, is governed by surface interactions. In addition, surface science is integral part of the formulation, manufacturing, and application of coatings.

Chemical Methods

A polymeric material can be functionalized by the addition of small moieties, oligomers, and even other polymers (grafting copolymers) onto the surface or interface.

Grafting Copolymers

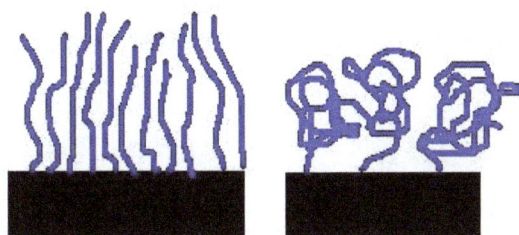

Grafting from a polymeric surface   **Grafting onto a polymeric surface**

The two methods of co-polymer grafting. Notice the difference in density of polymer chains, the equilibrium conformation of polymer molecules in solution gives the "mushroom" regime shown for the grafting-onto method.

Grafting, in the context of polymer chemistry, refers to the addition of polymer chains onto a surface. In the so-called 'grafting onto' mechanism, a polymer chain adsorbs onto a surface out of solution. In the more extensive 'grafting from' mechanism, a polymer chain is initiated and propagated at the surface. Because pre-polymerized chains used in the 'grafting onto' method have a thermodynamically favored conformation in solution (an equilibrium hydrodynamic volume), their adsorption density is self-limiting. The radius of gyration of the polymer therefore is the limiting factor in the number of polymer chains that can reach the surface and adhere. The 'grafting from' technique circumvents this phenomenon and allows for greater grafting densities.

The processes of grafting "onto", "from", and "through" are all different ways to alter the chemical reactivity of the surface they attach with. Grafting onto allows a preformed polymer, generally in a "mushroom regime", to adhere to the surface of either a droplet or bead in solution. Due to the larger volume of the coiled polymer and the steric hindrance this causes, the grafting density is lower for 'onto' in comparison to 'grafting from'. The surface of the bead is wetted by the polymer and the interaction in the solution caused the polymer to become more flexible. The 'extended conformation' of the polymer grafted, or polymerized, from the surface of the bead means that the monomer must be in the solution and there for lyophilic. This results with a polymer that has favorable interactions with the solution, allowing the polymer to form more linearly. Grafting from therefore has a higher grafting density since there are more access to chain ends.

Peptide synthesis can provide one example of a 'grafting from' synthetic process. In this process, an amio acid chain is grown by a series of condensation reaction from a polymer bead surface. This grafting technique allows for excellent control over the peptide composition as the bonded chain can be washed without desorption from the polymer.

Polymeric coatings are another area of applied grafting techniques. In the formulation of water-borne paint, latex particles are often surface modified to control particle dispersion and thus coating characteristics such as viscosity, film formation, and environmental stability (UV exposure and temperature variations).

Oxidation

Plasma processing, corona treatment, and flame treatment can all be classified as surface oxidation mechanisms. These methods all involve cleavage of polymer chains in the material and the incorporation of carbonyl, and hydroxyl functional groups. The incorporation of oxygen into the surface creates a higher surface energy allowing the substrate to be coated.

Methodology

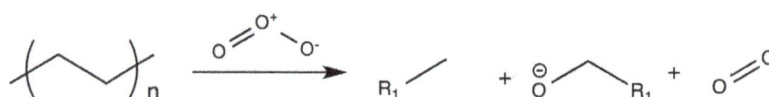

An example reaction scheme for the cleavage of bonds in the polymer chains of a polyolefin surface. The presence of ozone, as the result of an ionizing electric arc produced by a Corona treater for example, leads to oxidation of the surface yielding polar functionalities.

Oxidizing Polymeric Surfaces

Corona Treatment

Corona treatment is a surface modification method using a low temperature corona discharge to increase the surface energy of a material, often polymers and natural fibers. Most commonly, a thin polymer sheet is rolled through an array of high-voltage electrodes, using the plasma created to functionalize the surface. The limited penetration depth of such treatment provides vastly improved adhesion while preserving bulk mechanical properties.

Commercially, corona treatment has been used widely for improved dye adhesion before printing text and images on plastic packaging materials. The hazardous nature of remnant ozone after corona treatment stipulates careful filtration and ventilation during processing, restricting its implementation to applications with strict catalytic filtered systems. This limitation prevents widespread use within open-line manufacturing processes

Several factors influence the efficiency of the flame treatment such as air-to-gas ratio, thermal output, surface distance, and oxidation zone dwell time. Upon conception of the process, a corona treatment immediately followed film extrusions, but the development of careful transportation techniques allows treatment at an optimized location. Conversely, in-line corona treatments have been implemented into full-scale production lines such as those in the newspaper industry. These in-line solutions are developed to counteract the decrease in wetting characteristics caused by excessive solvent use.

Atmosphere- and Pressure-dependent Plasma Processing

Plasma processing provides interfacial energies and injected monomer fragments larger than comparable processes. However, limited fluxes prevent high process rates. In addition, plasmas are thermodynamically unfavorable and therefore plasma-processed surfaces lack uniformity, consistency, and permanence. These obstacles with plasma processing proclude it from being a competitive surface modification method within industry. The process begins with production of plasma via ionization either by deposition on monomer mixtures or gaseous carrier ions. The power required to produce the necessary plasma flux can be derived from the active volume mass/energy balance:

$$\int_{Vol_I} k^{ion} n_e n_0 \, dVol_I = \frac{n_e}{\tau_n} Vol_I$$

where

Vol_I is the active volume

k^{ion} is the ionization rate

n_0 is the neutral density

n_e is the electron density

τ_n is the ion loss by diffusion, convection, attachment, and recombination

Dissipation is generally initiated via direct current (DC), radio frequency (RF), or microwave power. Gas ionization efficiency can decrease the power efficiency more than tenfold depending on the carrier plasma and substrate.

Flamed Plasma Processing

Flame treatment is a controlled, rapid, cost-effective method of increasing surface energy and wettability of polyolefins and metallic components. This high-temperature plasma treatment uses ionized gaseous oxygen via jet flames across a surface to add polar functional groups while melting the surface molecules, locking them into place upon cooling.

Thermoplastic polyethylene and polypropylene treated with brief oxygen plasma exposure have seen contact angles as low as 22°, and the resulting surface modification can last years with proper packaging. Flame plasma treatment has become increasingly popular with intravascular devices such as balloon catheters due to the precision and cost-effectiveness demanded in the medical industry.

Grafting Techniques

Grafting copolymers to a surface can be envisioned as fixing polymeric chains to a structurally different polymer substrate with the intention of changing surface functionality while preserving bulk mechanical properties. The nature and degree of surface functionalization is determined by both the choice of copolymer and the type and extent of grafting.

Photografting

The modification of inert surfaces of polyolefins, polyesters, and polyamides by grafting functional vinyl monomers has been used to increase hydrophobicity, dye absorption, and polymer adhesion. This photografting method is generally used during continuous filament or thin film processing. On a bulk commercial scale, the grafting technique is referred to as photoinitiated lamination, where desired surfaces are joined by grafting a polymeric adhesion network between the two films. The low adhesion and absorption of polelfins, polesters, and polyamides is improved by UV-irradiation of an initiator and monomer transferred through the vapor phase to the substrate. Functionalization of porous surfaces have seen great success with high temperature photografting techniques.

In microfluidic chips, functionalizing channels allows directed flow to preserve lamellar behavior between and within junctions. The adverse turbulent flow in microfluidic applications can compound component failure modes due to the increased level of channel interdependency and network complexity. In addition, the imprinted design of microfluidic channels can be reproduced for photografting the corresponding channels with a high degree of accuracy.

Surface Analytical Techniques

Surface Energy Measurement

In industrial corona and plasma processes, cost-efficient and rapid analytical methods are required for confirming adequate surface functionality on a given substrate. Measuring the surface energy is an indirect method for confirming the presence of surface functional groups without the need for microscopy or spectroscopy, often expensive and demanding tools. Contact angle measurement (goniometry) can be used to find the surface energy of the treated and non-treated surface. Young's relation can be used to find surface energy assuming the simplification of experimental conditions to a three phase equilibrium (i.e. liquid drop applied to flat rigid solid surface in a controlled atmosphere), yielding

$$\gamma_{SG} = \gamma_{SL} + \gamma_{LG} \cos\theta_c$$

where

γ_{ij} denotes the surface energy of the solid–liquid, liquid–gas, or solid–gas interface

θ_c is the measured contact angle

A series of solutions with known surface tension (e.g., Dyne solutions) can be used to estimate the surface energy of the polymer substrate qualitatively by observing the wettability of each. These methods are applicable to macroscopic surface oxidation, as in industrial processing.

Infrared Spectroscopy

In the case of oxidizing treatments, spectra taken from treated surfaces will indicate the presence of functionalities in carbonyl and hydroxyl regions according to the Infrared spectroscopy correlation table.

XPS and EDS

X-ray photoelectron spectroscopy (XPS) and Energy-dispersive X-ray spectroscopy (EDS/EDX) are composition characterization techniques that use x-ray excitation of electrons to discrete energy levels to quantify chemical composition. These techniques provide characterization at surface depths of 1–10 nanometers, approximately the range of oxidation in plasma and corona treatments. In addition, these processes offer the benefit of characterizing microscopic variations in surface composition.

In the context of plasma processed polymer surfaces, oxidized surfaces will obviously show a greater oxygen content. Elemental analysis allows for quantitative data to be obtained and used in the analysis of process efficiency.

Atomic Force Microscopy

Atomic force microscopy (AFM), a type of scanning force microscopy, was developed for mapping three-dimensional topographical variations in atomic surfaces with high resolution (on the order of fraction of nanometers). AFM was developed to overcome the material conduction limitations of electron transmission and scanning microscopy methods (SEM & STM). Invented by Binnig, Quate, and Gerbe in 1985, atomic force microscopy uses laser beam deflection to measure the variations in atomic surfaces. The method does not rely on the variation in electron conduction through the material, as the scanning tunneling microscope (STM) does, and therefore allow microscopy on nearly all materials, including polymers.

The application of AFM on polymeric surfaces is especially favorable because polymer general lack of crystallinity leads to large variations in surface topography. Surface functionalization techniques such as grafting, corona treatment, and plasma processing increase the surface roughness greatly (compared to the unprocessed substrate surface) and are therefore accurately measured by AFM.

Applications

Biomaterials

Biomaterial surfaces are often modified using light-activated mechanisms (such as photografting) to functionalize the surface without compromising bulk mechanical properties.

The modification of surfaces to keep polymers biologically inert has found wide uses in biomedical applications such as cardiovascular stents and in many skeletal prostheses. Functionalizing polymer surfaces can inhibit protein adsorption, which may otherwise initiate cellular interrogation upon the implant, a predominant failure mode of medical prostheses.

Polymer	Medical Application	Functionalization Method & Purpose
Polyvinylchloride (PVC)	Endotracheal tubes	Plasma processed to increase hydrophobicity
Silicone rubber	Breast implants	Glow-discharge plasma processed coatings with halofuginone to prevent capsular fibrosis
Polyethylene (PE)	Synthetic vascular grafts	Polydimethylsiloxane (PDMS) microfluidic patterning for selective adsorption of fibronectin
Polymethylmethacrylate (PMMA)	Intraocular lenses	Photografting nanoelectromechanical structures to increase photopic sensitivity

Narrow biocompatibility requirements within the medical industry have over the past ten years driven surface modification techniques to reach an unprecedented level of accuracy.

Coatings

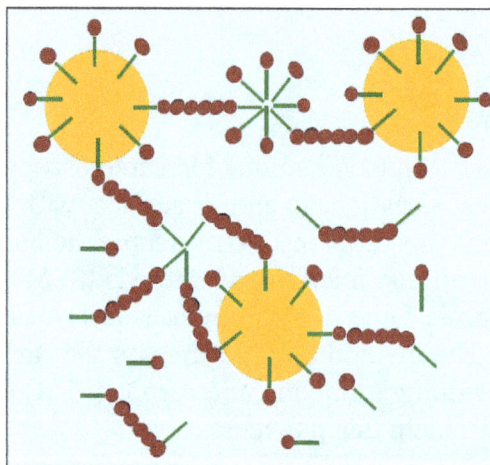

Adsorbed functionalities (e.g., surfact molecules) on a dispersed polymer particle interact with solvated associative thickeners (e.g., aqueous cellulosic polymer) yielding novel rheological behavior.

In water-borne coatings, an aqueous polymer dispersion creates a film on the substrate once the solvent has evaporated. Surface functionalization of the polymer particles is a key component of a coating formulation allowing control over such properties as dispersion, film formation temperature, and the coating rheology. Dispersing aids often involve steric or electrostatic repulsion of the polymer particles, providing colloidal stability. The dispersing aids adsorb (as in a grafting onto scheme) onto latex particles giving them functionality. The association of other additives, such as thickeners shown in the schematic to the right, with adsorbed polymer material give rise to complex rheological behavior and excellent control over a coating's flow properties.

Two-dimensional Polymer

A two-dimensional polymer (2DP) is a sheet-like monomolecular macromolecule consisting of laterally connected repeat units with end groups along all edges. This recent definition of 2DP is based on Hermann Staudinger's polymer concept from the 1920s. According to this, covalent long chain molecules ("Makromoleküle") do exist and are composed of a sequence of linearly connected repeat units and end groups at both termini.

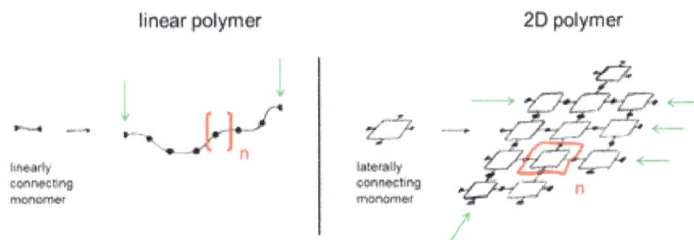

Figure 1. Structural difference between a linear and a two-dimensional (2D) polymer. In the former, linearly connecting monomers result in a thread-like linear polymer, while in the latter laterally connecting monomers result in a sheet-like 2DP with regularly tessellated repeat units (here of square geometry). The repeat units are marked in red, whereby the number n describes the degree of polymerization. While a linear polymer has two end groups, a 2DP has an infinite number of end groups that are positioned all along the sheet edges (green arrows).

Moving from one dimension to two offers access to surface morphologies such as increased surface area, porous membranes, and possibly in-plane pi orbital-conjugation for enhanced electronic properties. They are distinct from other families of polymers because 2D polymers can be isolated as multilayer crystals or as individual sheets.

The term 2D polymer has also been used more broadly to include linear polymerizations performed at interfaces, layered non-covalent assemblies, or to irregularly cross-linked polymers confined to surfaces or layered films. 2D polymers can be organized based on these methods of linking (monomer interaction): covalently linked monomers, coordination polymers and supramolecular polymers.

Topologically, 2DPs may thus be understood as structures made up from regularly tessellated regular polygons (the repeat units). Figure 1 displays the key features of a linear and a 2DP according to this definition. For usage of the term "2D polymer" in a wider sense.

Covalently-Linked Polymers

There are several examples of covalently linked 2DPs which include the individual layers or sheets of graphite (called graphenes), MoS2, (BN)x and layered covalent organic frameworks. As required by the above definition, these sheets have a periodic internal structure.

i. Graphene: A well-known example of a 2D polymer is graphene; whose optical, electronic and mechanical properties have been studied in depth. Graphene has a honeycomb lattice of carbon atoms that exhibit semiconducting properties. A potential repeat unit of graphene is a sp2-hybridized carbon atom. Individual sheets can in principle be obtained by exfoliation procedures, though in reality this is a non-trivial enterprise.

ii. MoS_2: Molybdenumdisulfide can exist in two-dimensional, single or layered polymers where each Mo(IV) center occupies a trigonal prismatic coordination sphere.

iii. BN: Boron nitride polymers are stable in its crystalline hexagonal form where it has a two-dimensional layered structure similar to graphene. There are covalent bonds formed between boron and nitrogen atoms, yet the layers are held together by weak van der Waals interactions, in which the boron atoms are eclipsed over the nitrogen.

iv. Covalent Organic Frameworks

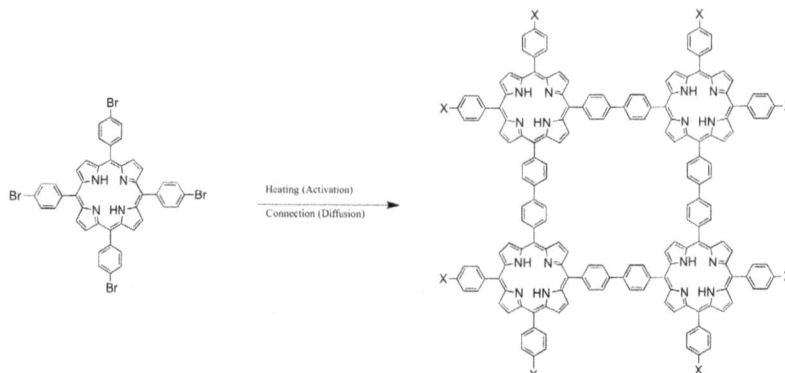

Figure 2. Surface-mediated 2D polymerization scheme of the tetrafunctional porphyrin monomer.

Two dimensional covalent organic frameworks (COFs) are one type of microporous coordination polymer that can be fabricated in the 2D plane. The dimensionality and topology of the 2D COFs result from both the shape of the monomers and the relative and dimensional orientations of their reactive groups. These materials contain desirable properties in fields of materials chemistry including thermal stability, tunable porosity, high specific surface area, and the low density of organic material. By careful selection of organic building units, long range π-orbital overlap parallel to the stacking direction of certain organic frameworks can be achieved.

Figure 3. 2D polymerization under thermodynamic control (top) versus kinetic control (bottom). Solid black lines represent covalent bond formation

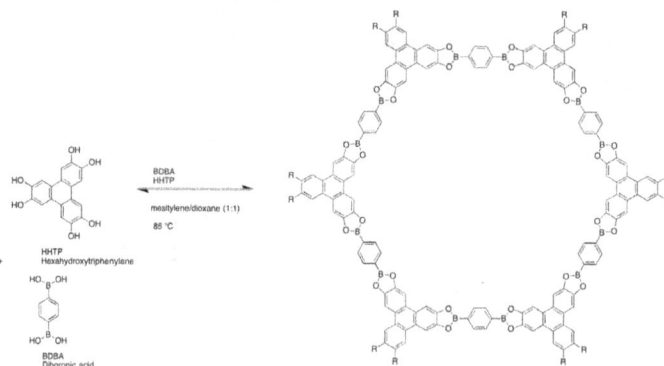

Figure 4. Synthetic scheme of covalent organic framework using Boronic acid and hexahydroxytriphenylene (HHTP).

Many covalent organic frameworks derive their topology from the directionality of the covalent linkages, thus small changes in organic linkers can dramatically affect their mechanical and electronic properties. Even small changes in their structure can induce dramatic changes in stacking behavior of molecular semiconductors.

Porphyrins are an additional class of conjugated, heterocyclic macrocycles. Control of monomer assembly through covalent assembly has also been demonstrated using covalent interactions with porphyrins. Upon thermal activation of porphyrin building blocks, covalent bonds form to create a conductive polymer, a versatile route for bottom-up construction of electronic circuits been demonstrated. (Figure 2)

Synthesis

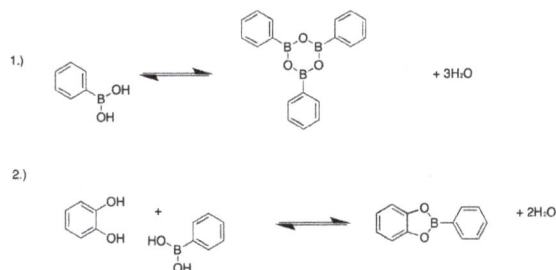

Figure 5. Boronate ester equilibria used to prepare various 2D COFs

It is possible to synthesize COFs using both dynamic covalent and non-covalent chemistry. The kinetic approach involves a stepwise process of polymerizing pre-assembled 2D-monomer while thermodynamic control exploits reversible covalent chemistry to allow simultaneous monomer assembly and polymerization. Under thermodynamic control, bond formation and crystallization also occur simultaneously. (Figure 3) Covalent organic frameworks formed by dynamic covalent bond formation involves chemical reactions carried out reversibly under conditions of equilibrium control. Because the formation of COFs in dynamic covalent formation occurs under thermodynamic control, product distributions depend only on the relative stabilities of the final products. (Figure 3) Covalent assembly to form 2D COFs has been previously done using boronate esters from catechol acetonides in the presence of a lewis acid (BF_3*OEt_2). (Figure 4)

2D polymerization under kinetic control relies on non-covalent interactions and monomer assembly prior to bond formation. The monomers can be held together in a pre-organized position by non-covalent interactions, such as hydrogen bonding or van der Waals.

Coordination Polymers

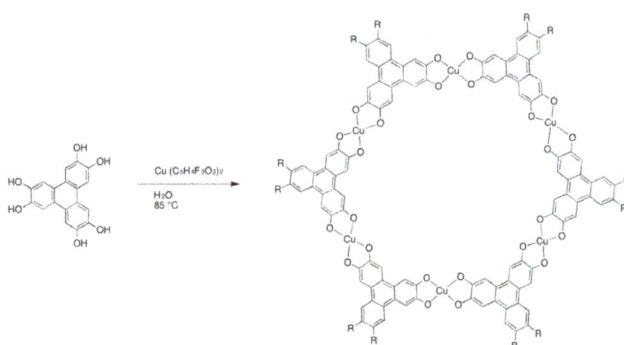

Figure 6. Synthetic scheme for metal organic framework (MOF) using hexahydroxytriphenylene (HHTP) and Cu(II) metal.

Metal Organic Frameworks

Self-assembly can also be observed in the presence of organic ligands and various metals centers through coordinative bonds or supramolecular interactions. Molecular self- assembly involves the association by many weak, reversible interactions to obtain a final structure that represents a thermodynamic minimum. A class of coordination polymers, known also as metal-organic frameworks (MOFs), are metal-ligand compounds that extend "infinitely" into one, two or three dimensions.

Synthesis

Availability of modular metal centers and organic building blocks generate wide diversity in synthetic versatility. Their applications range from industrial use to chemiresistive sensors. The ordered structure of the frame is largely determined by the coordination geometry of the metal and directionality of functional groups upon the organic linker. Consequently, MOFs contain highly defined pore dimensions when compared with conventional amorphous nanoporous materials and polymers. Reticular Synthesis of MOFs is a term that has been recently coined to describe the bottom-up method of assembling cautiously designed rigid molecular building blocks into pre-arranged structures held together by strong chemical bonds. The synthesis of two-dimensional MOFs begins with the knowledge of a target "blueprint" or a network, followed by identification of the required building blocks for its assembly.

By interchanging metal centers and organic ligands, one can fine-tune electronic and magnetic properties observed in MOFs. There have been recent efforts synthesize conductive MOFs using triyphenylene linkers (Figure 6). Additionally, MOFs have been utilized as reversible chemiresistive sensors.

Supramolecular Polymers

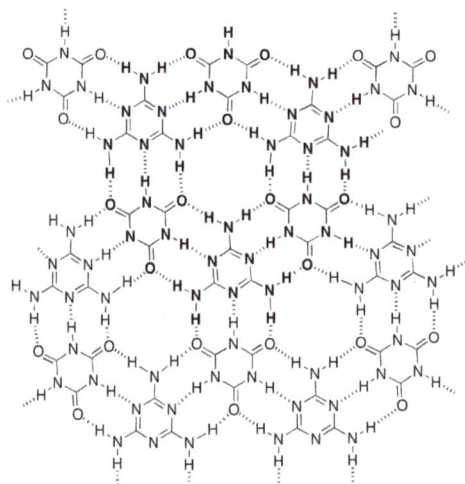

Figure 7. Supramolecular aggregates of (CA*M) cyanuric acid (CA) and melamine (M).

Supramolecular assembly requires non-covalent interactions directing the formation of 2D polymers by relying on electrostatic interactions such as hydrogen bonding and van der Waals forces. To design artificial assemblies capable of high selectivity requires correct manipulation of energetic and stereochemical features of non-covalent forces. Some benefits of non-covalent interactions is their reversible nature and response to external factors such as temperature and concentration. The mechanism of non-covalent polymerization in supramolecular chemistry is highly dependent on the interactions during the self-assembly process. The degree of polymerization depends highly on temperature and concentration. The mechanisms may be divided into three categories: isodesmic, ring-chain, and cooperative.

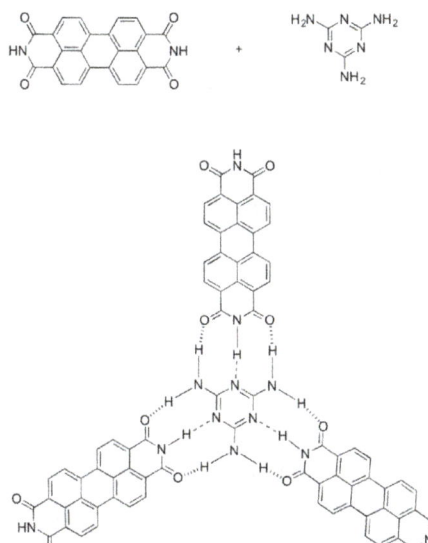

Figure 8. Self-assembly of a PTCDI–melamine supramolecular network. Dotted lines represent the stabilizing hydrogen bonds between the molecules.

One example of isodesmic associations in supramolecular aggregates is seen in Figure 7, (CA*M) cyanuric acid (CA) and melamine (M) interactions and assembly through hydrogen bonding. Hydrogen bonding has been used to guide assembly of molecules into two-dimensional networks, that can then serve as new surface templates and offer an array of pores of sufficient capacity to accommodate large guest molecules. (Figure 7) An example of utilizing surface structures through non-covalent assembly uses adsorbed monolayers to create binding sites for target molecules through hydrogen bonding interactions. Hydrogen bonding is used to guide the assembly of two different molecules into a 2D honeycomb porous network under ultra high vacuum seen in figure 8.

Characterization

2DPs as two dimensional sheet macromolecules have a crystal lattice, that is they consist of monomer units that repeat in two dimensions. Therefore, a clear diffraction pattern from their crystal lattice should be observed as a proof of crystallinity. The internal periodicity is supported by electron microscopy imaging, electron diffraction and Raman-spectroscopic analysis.

2DPs should in principle also be obtainable by, e.g., an interfacial approach whereby proving the internal structure, however, is more challenging and has not yet been achieved.

In 2014 a 2DP was reported synthesised from a trifunctional photoreactive anthracene derived monomer, preorganised in a lamellar crystal and photopolymerised in a [4+4]cycloaddition. Another reported 2DP also involved an anthracene-derived monomer

Applications

2DPs are expected to be superb membrane materials because of their defined pore sizes. Furthermore, they can serve as ultrasensitive pressure sensors, as precisely defined catalyst supports, for surface coatings and patterning, as ultrathin support for cryo-TEM, and many other applications.

Since 2D polymers provide an availability of large surface area and uniformity in sheets, they also found useful applications in areas such as selective gas adsorption and separation. Metal organic frameworks have become popular recently due to the variability of structures and topology which provide tunable pore structures and electronic properties. There are also ongoing methods for creation of nanocrystals of MOFs and their incorporation into nanodevices. Additionally, metal-organic surfaces have been synthesized with cobalt dithionlene catalysts for efficient hydrogen production through reduction of water as an important strategy for fields of renewable energy.

The fabrication of 2D organic frameworks, have also synthesized two-dimensional, porous covalent organic frameworks to be used as storage media for hydrogen, methane and carbon dioxide in clean energy applications.

History

First attempts to synthesize 2DPs date back to the 1930s when Gee reported interfacial polymerizations at the air/water interface in which a monolayer of an unsaturated fatty acid derivative was laterally polymerized to give a 2D cross-linked material. Since then a number of important attempts were reported in terms of cross-linking polymerization of monomers confined to layered templates or various interfaces. These approaches provide easy accesses to sheet-like polymers. However, the sheets' internal network structures are intrinsically irregular and the term "repeat unit" is not applicable. In organic chemistry, creation of 2D periodic network structures has been a dream for decades. Another noteworthy approach is "on-surface polymerization" whereby 2DPs with lateral dimensions not exceeding some tens of nanometers were reported. Laminar crystals are readily available, each layer of which can ideally be regarded as latent 2DP. There have been a number of attempts to isolate the individual layers by exfoliation techniques.

Tacticity

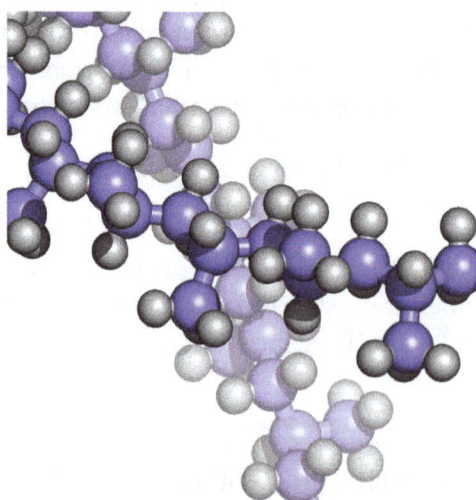

A ball-and-stick model of syndiotactic polypropylene.

Tacticity is the relative stereochemistry of adjacent chiral centers within a macromolecule. The practical significance of tacticity rests on the effects on the physical properties of the polymer. The regularity of the macromolecular structure influences the degree to which it has rigid, crystalline long range order or flexible, amorphous long range disorder. Precise knowledge of tacticity of a polymer also helps understanding at what temperature a polymer melts, how soluble it is in a solvent and its mechanical properties.

A tactic macromolecule in the IUPAC definition is a macromolecule in which essentially all the configurational (repeating) units are identical. Tacticity is particularly significant in vinyl polymers of the type -H_2C-CH(R)- where each repeating unit with a substituent R on one side of the polymer backbone is followed by the next repeating unit with the substituent on the same side as the previous one, the other side as the previous one or positioned randomly with respect to the previous one. In a hydrocarbon macromolecule with all carbon atoms making up the backbone in a tetrahedral molecular geometry, the zigzag backbone is in the paper plane with the substituents either sticking out of the paper or retreating into the paper. This projection is called the Natta projection after Giulio Natta. Monotactic macromolecules have one stereoisomeric atom per repeat unit, ditactic to n-tactic macromolecules have more than one stereoisomeric atom per unit.

IUPAC Definition

The orderliness of the succession of configurational repeating units in the main chain of a regular macromolecule, a regular oligomer molecule, a regular block, or a regular chain.

Describing Tacticity

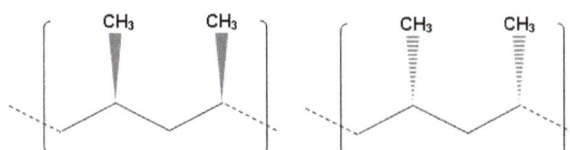

An example of *meso* diads in a polypropylene molecule.

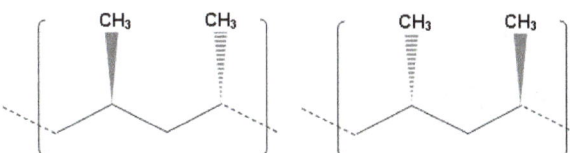

An example of *racemo* diads in a polypropylene molecule.

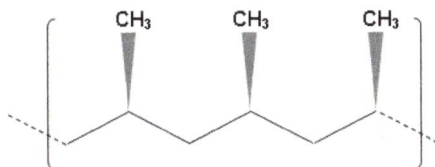

An isotactic (*mm*) triad in a polypropylene molecule.

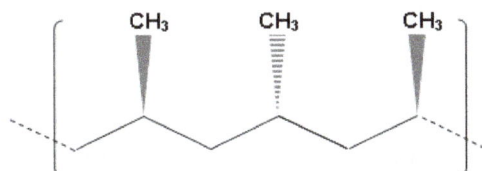

A syndiotactic (*rr*) triad in a polypropylene molecule.

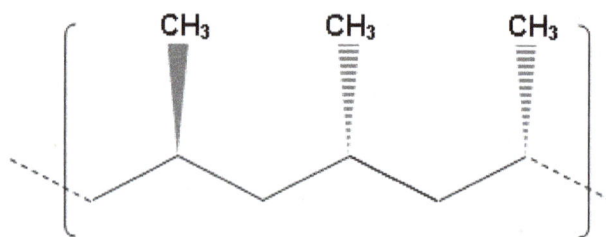

A heterotactic (*rm*) triad in a polypropylene molecule.

Diads

Two adjacent structural units in a polymer molecule constitute a diad. If the diad consists of two identically oriented units, the diad is called a meso diad reflecting similar features as a meso compound. If the diad consists of units oriented in opposition, the diad is called a racemo diad as in a racemic compound. In the case of vinyl polymer molecules, a meso diad is one in which the book carbon chains are oriented on the same side of the polymer backbone.

Triads

The stereochemistry of macromolecules can be defined even more precisely with the introduction of triads. An isotactic triad (mm) is made up of two adjacent meso diads, a syndiotactic triad {rr} consists of two adjacent racemo diads and a heterotactic triad (rm) is composed of a meso diad adjacent to a racemo diad. The mass fraction of isotactic (mm) triads is a common quantitative measure of tacticity.

When the stereochemistry of a macromolecule is considered to be a Bernoulli process, triad composition can be calculated from the probability of finding meso diads (P_m). When this probability is 0.25 then the probability of finding:

- an isotactic triad is P_m^2 or 0.0625

- an heterotactic triad is $2P_m(1-P_m)$ or 0.375

- a syndiotactic triad is $(1-P_m)^2$ or 0.5625

with a total probability of 1. Similar relationships with diads exist for tetrads.

Tetrads, Pentads, etc.

The definition of tetrads and pentads introduce further sophistication and precision to defining tacticity, especially when information on long-range ordering is desirable. Tacticity measurements obtained by Carbon-13 NMR are typically expressed in terms of the relative abundance of various pentads within the polymer molecule, e.g. *mmmm, mrrm*.

Other Conventions for Quantifying Tacticity

The primary convention for expressing tacticity is in terms of the relative weight fraction of triad or higher-order components, as described above. An alternative expression for tacticity is the average length of *meso* and *racemo* sequences within the polymer molecule. The average meso sequence length may be approximated from the relative abundance of pentads as follows:

$$MSL = \frac{mmmm + \frac{3}{2}mrrr + 2rmmr + \frac{1}{2}rmrm + \frac{1}{2}rmrr}{\frac{1}{2}mmmr + rmmr + \frac{1}{2}rmrm + \frac{1}{2}rmrr}$$

Polymers

Isotactic Polymers

Isotactic polymers are composed of isotactic macromolecules (IUPAC definition). In isotactic macromolecules all the substituents are located on the same side of the macromolecular backbone. An isotactic macromolecule consists of 100% meso diads. Polypropylene formed by Ziegler-Natta catalysis is an isotactic polymer. Isotactic polymers are usually semicrystalline and often form a helix configuration.

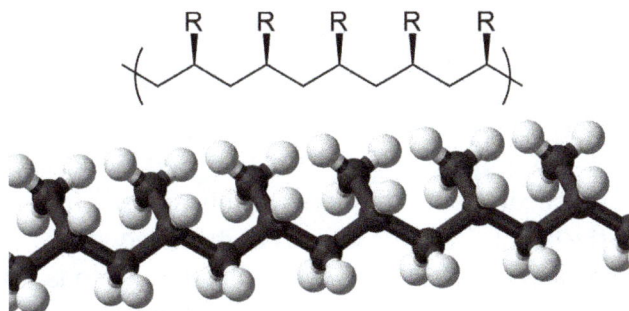

Syndiotactic Polymers

In syndiotactic or syntactic macromolecules the substituents have alternate positions along the chain. The macromolecule consists 100% of racemo diads. Syndiotactic polystyrene, made by metallocene catalysis polymerization, is crystalline with a melting point of 161 °C. Gutta percha is also an example for Syndiotactic polymer.

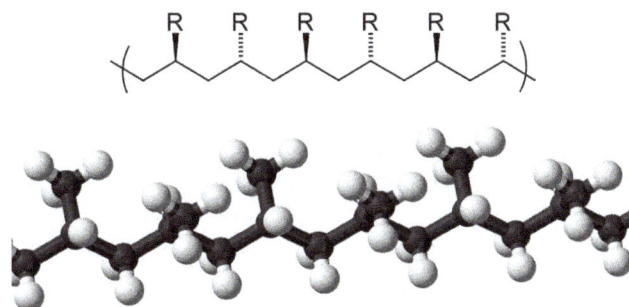

Atactic Polymers

In atactic macromolecules the substituents are placed randomly along the chain. The percentage of meso diads is between 1 and 99%. With the aid of spectroscopic techniques such as

NMR it is possible to pinpoint the composition of a polymer in terms of the percentages for each triad.

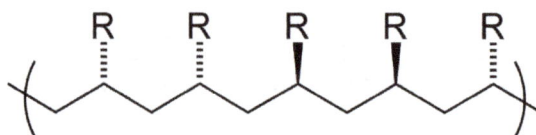

Polymers that are formed by free-radical mechanisms such as polyvinyl chloride are usually atactic. Due to their random nature atactic polymers are usually amorphous. In hemi isotactic macromolecules every other repeat unit has a random substituent.

Atactic polymers are technologically very important. A good example is polystyrene (PS). If a special catalyst is used in its synthesis it is possible to obtain the syndiotactic version of this polymer, but most industrial polystyrene produced is atactic. The two materials have very different properties because the irregular structure of the atactic version makes it impossible for the polymer chains to stack in a regular fashion. The result is that, whereas syndiotactic PS is a semicrystalline material, the more common atactic version cannot crystallize and forms a *glass* instead. This example is quite general in that many polymers of economic importance are atactic glass formers.

Eutactic Polymers

In eutactic macromolecules, substituents may occupy any specific (but potentially complex) sequence of positions along the chain. Isotactic and syndiotactic polymers are instances of the more general class of eutactic polymers, which also includes heterogeneous macromolecules in which the sequence consists of substituents of different kinds (for example, the side-chains in proteins and the bases in nucleic acids).

Head/Tail Configuration

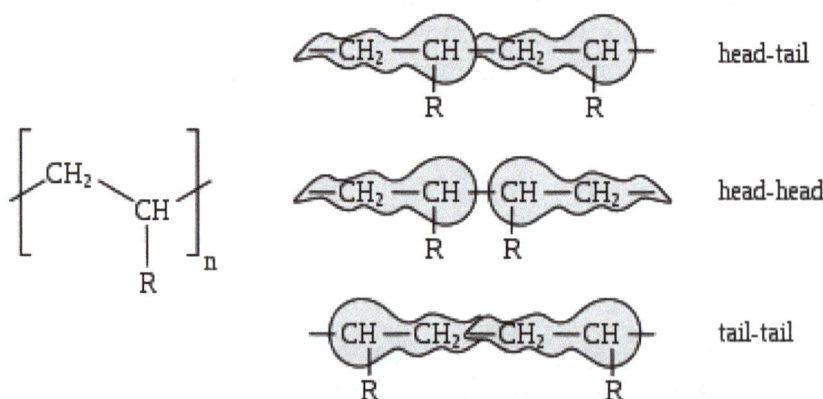

In vinyl polymers the complete configuration can be further described by defining polymer head/tail configuration. In a regular macromolecule all monomer units are normally linked in a head to tail configuration so that all β-substituents are separated by three carbon atoms. In head to head configuration this separation is only by 2 carbon atoms and the separation with tail to tail configuration is by 4 atoms. Head/tail configurations are not part of polymer tacticity but should be taken into account when considering polymer defects.

Techniques for Measuring Tacticity

Tacticity may be measured directly using proton or carbon-13 NMR. This technique enables quantification of the tacticity distribution by comparison of peak areas or integral ranges corresponding to known diads (r, m), triads (mm, rm+mr, rr) and/or higher order n-ads depending on spectral resolution. In cases of limited resolution stochastic methods such as Bernoullian or Markovian analysis may also be used to fit the distribution and back then forward predict higher n-ads and calculate the isotacticity of the polymer to the desired level.

Other techniques sensitive to tacticity include x-ray powder diffraction, secondary ion mass spectrometry (SIMS), vibrational spectroscopy (FTIR) and especially two-dimensional techniques. Tacticity may also be inferred by measuring another physical property, such as melting temperature, when the relationship between tacticity and that property is well-established.

Ideal Chain

An ideal chain (or freely-jointed chain) is the simplest model to describe polymers, such as nucleic acids and proteins. It only assumes a polymer as a random walk and neglects any kind of interactions among monomers. Although it is simple, its generality gives insight about the physics of polymers.

In this model, monomers are rigid rods of a fixed length l, and their orientation is completely independent of the orientations and positions of neighbouring monomers, to the extent that two monomers can co-exist at the same place. In some cases, the monomer has a physical interpretation, such as an amino acid in a polypeptide. In other cases, a monomer is simply a segment of the polymer that can be modeled as behaving as a discrete, freely jointed unit. If so, l is the Kuhn length. For example, chromatin is modeled as a polymer in which each monomer is a segment approximately 14-46 kbp in length.

The Model

N monomers form the polymer, whose total unfolded length is:

$L = Nl$, where N is the number of monomers.

In this very simple approach where no interactions between monomers are considered, the energy of the polymer is taken to be independent of its shape, which means that at thermodynamic equilibrium, all of its shape configurations are equally likely to occur as the polymer fluctuates in time, according to the Maxwell–Boltzmann distribution.

Let us call $\vec{R}$ the total end to end vector of an ideal chain and $\vec{r}_1, \dots, \vec{r}_N$ the vectors corresponding to individual monomers. Those random vectors have components in the three directions of space. Most of the expressions given in this article assume that the number of monomers N is large, so that the central limit theorem applies. The figure below shows a sketch of a (short) ideal chain.

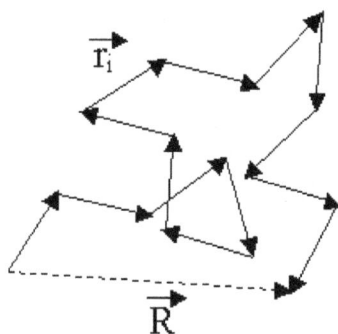

The two ends of the chain are not coincident, but they fluctuate around each other, so that of course:

$$\langle \vec{R} \rangle = \Sigma_{i=1}^{N} \langle \vec{r_i} \rangle = \vec{0}$$

Throughout the article the $\langle \rangle$ brackets will be used to denote the mean (of values taken over time) of a random variable or a random vector, as above.

Since $\vec{r}_1, \ldots, \vec{r}_N$ are independent, it follows from the Central limit theorem that $\vec{R}$ is distributed according to a normal distribution (or gaussian distribution): precisely, in 3D, R_x, R_y, and R_z are distributed according to a normal distribution of mean o and of variance:

$$\sigma^2 = \langle R_x^2 \rangle - \langle R_x \rangle^2 = \langle R_x^2 \rangle - 0$$

$$\langle R_x^2 \rangle = \langle R_y^2 \rangle = \langle R_z^2 \rangle = N\frac{l^2}{3}$$

So that $\langle \overrightarrow{R^2} \rangle = Nl^2 = Ll$. The end to end vector of the chain is distributed according to the following probability density function:

$$P(\vec{R}) = \left(\frac{3}{2\pi Nl^2} \right)^{3/2} e^{-\frac{3\vec{R}^2}{2Nl^2}}$$

The average end-to-end distance of the polymer is:

$$\sqrt{\langle \overrightarrow{R^2} \rangle} = \sqrt{N}l = \sqrt{Ll}$$

A quantity frequently used in polymer physics is the radius of gyration:

$$R_G = \frac{\sqrt{N}l}{\sqrt{6}}$$

It is worth noting that the above average end-to-end distance, which in the case of this simple model is also the typical amplitude of the system's fluctuations, becomes negligible compared to the total unfolded length of the polymer Nl at the thermodynamic limit. This result is a general property of statistical systems.

Mathematical remark: the rigorous demonstration of the expression of the density of probability $P(\vec{R})$ is not as direct as it appears above: from the application of the usual (1D) central limit theorem one can deduce that R_x, R_y and R_z are distributed according to a centered normal distribution of variance $Nl^2/3$. Then, the expression given above for $P(\vec{R})$ is not the only one that is compatible with such distribution for R_x, R_y and R_z. However, since the components of the vectors $\vec{r}_1,...,\vec{r}_N$ are uncorrelated for the random walk we are considering, it follows that R_x, R_y and R_z are also uncorrelated. This additional condition can only be fulfilled if $\vec{R}$ is distributed according to $P(\vec{R})$. Alternatively, this result can also be demonstrated by applying a multidimensional generalization of the central limit theorem, or through symmetry arguments.

Generality of the Model

While the elementary model described above is totally unadapted to the description of real-world polymers at the microscopic scale, it does show some relevance at the macroscopic scale in the case of a polymer in solution whose monomers form an ideal mix with the solvent (in which case, the interactions between monomer and monomer, solvent molecule and solvent molecule, and between monomer and solvent are identical, and the system's energy can be considered constant, validating the hypotheses of the model).

The relevancy of the model is, however, limited, even at the macroscopic scale, by the fact that it does not consider any excluded volume for monomers (or, to speak in chemical terms, that it neglects steric effects).

Other fluctuating polymer models that consider no interaction between monomers and no excluded volume, like the worm-like chain model, are all asymptotically convergent toward this model at the thermodynamic limit. For purpose of this analogy a Kuhn segment is introduced, corresponding to the equivalent monomer length to be considered in the analogous ideal chain. The number of Kuhn segments to be considered in the analogous ideal chain is equal to the total unfolded length of the polymer divided by the length of a Kuhn segment.

Entropic Elasticity of an Ideal Chain

If the two free ends of an ideal chain are attached to some kind of micro-manipulation device, then the device experiences a force exerted by the polymer. The ideal chain's energy is constant, and thus its time-average, the internal energy, is also constant, which means that this force necessarily stems from a purely entropic effect.

This entropic force is very similar to the pressure experienced by the walls of a box containing an ideal gas. The internal energy of an ideal gas depends only on its temperature, and not on the volume of its containing box, so it is not an energy effect that tends to increase the volume of the box like gas pressure does. This implies that the pressure of an ideal gas has a purely entropic origin.

What is the microscopic origin of such an entropic force or pressure? The most general answer is that the effect of thermal fluctuations tends to bring a thermodynamic system toward a macroscopic state that corresponds to a maximum in the number of microscopic states (or micro-states) that are compatible with this macroscopic state. In other words, thermal fluctuations tend to bring a system toward its macroscopic state of maximum entropy.

What does this mean in the case of the ideal chain? First, for our ideal chain, a microscopic state is characterized by the superposition of the states $\vec{r}_i$ of each individual monomer (with i varying from 1 to N). In its solvent, the ideal chain is constantly subject to shocks from moving solvent molecules, and each of these shocks sends the system from its current microscopic state to another, very similar microscopic state. For an ideal polymer, as will be shown below, there are more microscopic states compatible with a short end-to-end distance than there are microscopic states compatible with a large end-to-end distance. Thus, for an ideal chain, maximizing its entropy means reducing the distance between its two free ends. Consequently, a force that tends to collapse the chain is exerted by the ideal chain between its two free ends.

In this section, the mean of this force will be derived. The generality of the expression obtained at the thermodynamic limit will then be discussed.

Ideal Chain Under Length Constraint

The case of an ideal chain whose two ends are attached to fixed points will be considered in this sub-section. The vector $\vec{R}$ joining these two points characterizes the macroscopic state (or macro-state) of the ideal chain. To each macro-state corresponds a certain number of micro-states, that we will call $\Omega(\vec{R})$ (micro-states are defined in the introduction to this section). Since the ideal chain's energy is constant, each of these micro-states is equally likely to occur. The entropy associated to a macro-state is thus equal to:

$$S(\vec{R}) = k \ \log(\Omega(\vec{R})) \text{, where } k_B \text{ is Boltzmann's constant}$$

The above expression gives the absolute (quantum) entropy of the system. A precise determination of $\Omega(\vec{R})$ would require a quantum model for the ideal chain, which is beyond the scope of this article. However, we have already calculated the probability density $P(\vec{R})$ associated with the end-to-end vector of the *unconstrained* ideal chain, above. Since all micro-states of the ideal chain are equally likely to occur, $P(\vec{R})$ is proportional to $\Omega(\vec{R})$. This leads to the following expression for the classical (relative) entropy of the ideal chain:

$$S(\vec{R}) = k_B \log(P(\vec{R})) + C_{st},$$

where C_{st} is a fixed constant. Let us call $\vec{F}$ the force exerted by the chain on the point to which its end is attached. From the above expression of the entropy, we can deduce an expression of this force. Suppose that, instead of being fixed, the positions of the two ends of the ideal chain are now controlled by an operator. The operator controls the evolution of the end to end vector $\vec{R}$. If the operator changes $\vec{R}$ by a tiny amount $\overrightarrow{dR}$, then the variation of internal energy of the chain is zero, since the energy of the chain is constant. This condition can be written as:

$$0 = dU = \delta W + \delta Q$$

δW is defined as the elementary amount of mechanical work transferred by the operator to the ideal chain, and δQ is defined as the elementary amount of heat transferred by the solvent to the ideal chain. Now, if we assume that the transformation imposed by the operator on the system is quasistatic (i.e., infinitely slow), then the system's transformation will be time-reversible, and we can assume that during its passage from macro-state $\vec{R}$ to macro-state $\vec{R} + \overrightarrow{dR}$, the system passes through a series of thermodynamic equilibrium macro-states. This has two consequences:

- first, the amount of heat received by the system during the transformation can be tied to the variation of its entropy:

$\delta Q = T dS$, where T is the temperature of the chain.

- second, in order for the transformation to remain infinitely slow, the mean force exerted by the operator on the end points of the chain must balance the mean force exerted by the chain on its end points. Calling $\vec{f}_{op}$ the force exerted by the operator and $\vec{f}$ the force exerted by the chain, we have:

$$\delta W = \langle \vec{f}_{op} \rangle \cdot \overrightarrow{dR} = -\langle \vec{f} \rangle \cdot \overrightarrow{dR}$$

We are thus led to:

$$\langle \vec{f} \rangle = T \frac{dS}{\overrightarrow{dR}} = \frac{k_B T}{P(\vec{R})} \frac{dP(\vec{R})}{\overrightarrow{dR}}$$

$$\langle \vec{f} \rangle = -k_B T \frac{3\vec{R}}{Nl^2}$$

The above equation is the equation of state of the ideal chain. Since the expression depends on the central limit theorem, it is only exact in the limit of polymers containing a large number of monomers (that is, the thermodynamic limit). It is also only valid for small end-to-end distances, relative to the overall polymer contour length, where the behavior is like a hookean spring. Behavior over larger force ranges can be modeled using a canonical ensemble treatment identical to magnetization of paramagnetic spins. For the arbitrary forces the extension-force dependence will be given by Langevin function:

$$\frac{R}{Nl} = \coth\left(\frac{fl}{k_B T}\right) - \frac{k_B T}{fl}$$

For the arbitrary extensions the force-extension dependence can be approximated by:

$$\frac{fl}{k_B T} = 3\frac{R}{Nl} + \frac{1}{5}\left(\frac{R}{Nl}\right)^2 \sin\left(\frac{7R}{2Nl}\right) + \frac{\left(\frac{R}{Nl}\right)^3}{1 \quad R}$$

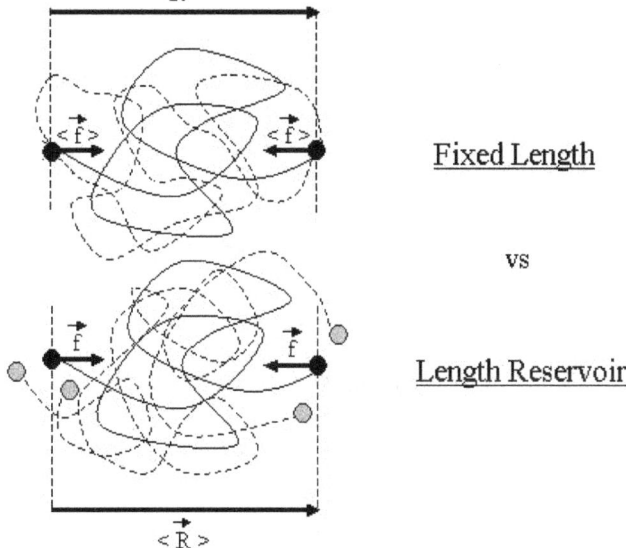

Fixed Length

vs

Length Reservoir

Finally, the model can be extended to even larger force ranges by inclusion of a stretch modulus along the polymer contour length. That is, by allowing the length of each unit of the chain to respond elastically to the applied force.

Ideal Polymer Exchanging Length with a Reservoir

Throughout this sub-section, as in the previous one, the two ends of the polymer are attached to a micro-manipulation device. This time, however, the device does not maintain the two ends of the ideal chain in a fixed position, but rather it maintains a constant pulling force $\vec{f}_{op}$ on the ideal chain. In this case the two ends of the polymer fluctuate around a mean position $\langle \vec{R} \rangle$. The ideal chain reacts with a constant opposite force $\vec{f} = -\vec{f}_{op}$

For an ideal chain exchanging length with a reservoir, a macro-state of the system is characterized by the vector $\vec{f}$.

The change between an ideal chain of fixed length and an ideal chain in contact with a length reservoir is very much akin to the change between the micro-canonical ensemble and the canonical ensemble. The change is from a state where a fixed value is imposed on a certain parameter, to a state where the system is left free to exchange this parameter with the outside. The parameter in question is energy for the microcanonical and canonical descriptions, whereas in the case of the ideal chain the parameter is the length of the ideal chain.

As in the micro-canonical and canonical ensembles, the two descriptions of the ideal chain differ only in the way they treat the system's fluctuations. They are thus equivalent at the thermodynamic limit. The equation of state of the ideal chain remains the same, except that is now subject to fluctuations:

$$\vec{f} = -k_B T \frac{3 \langle \vec{R} \rangle}{Nl^2} \ .$$

End-group

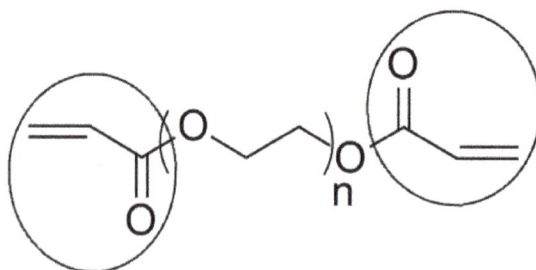

End group example of poly(ethylene glycol) diacrylate with the end groups circled

End groups are an important aspect of polymer synthesis and characterization. In polymer chemistry, end groups are functionalities or constitutional units that are at the extremity of a macromolecule or oligomer (IUPAC). In polymer synthesis, like condensation polymerization and free-radical types of polymerization, end-groups are commonly used and can be analyzed for example by nuclear magnetic resonance (NMR) to determine the average length of the polymer. Other methods for characterization of polymers where end-groups are used are mass spectrometry and vibrational spectrometry, like infrared and Raman spectrometry. Not only are these groups important for the analysis of the polymer, but they are also useful for grafting to and from a polymer chain to create a new copolymer. One example of an end group is in the polymer poly(ethylene glycol) diacrylate where the end-groups are circled.

End Groups in Polymer Synthesis

End groups are seen on all polymers and the functionality of those end groups can be important in determining the application of polymers. Each type of polymerization (free radical, condensation or etc.) have end groups that are typical for the polymerization, and knowledge of these can help to identify the type of polymerization method used to form the polymer.

Step-growth Polymerization

Step-growth polymerization involves two monomers with bi- or multifunctionality to form polymer chains. Many polymers are synthesized via step-growth polymerization and include polyesters, polyamides, and polyurethanes. A sub class of step-growth polymerization is condensation polymerization.

Condensation Polymerization

Condensation polymerization is an important class of step-growth polymerization, which is formed simply by the reaction of two monomers and results in the release of a water molecule. Since these polymers are typically made up of two or more monomers, the resulting end groups are from the monomer functionality. Examples of condensation polymers can be seen with polyamides, polyacetals and polyesters. An example of polyester is polyethylene terephthalate (PET), which is made from the monomers terephthalic acid and ethylene glycol. If one of the components in the polymerization is in excess, then that polymers functionality will be at the ends of the polymers (a carboxylic acid or alcohol group respectively).

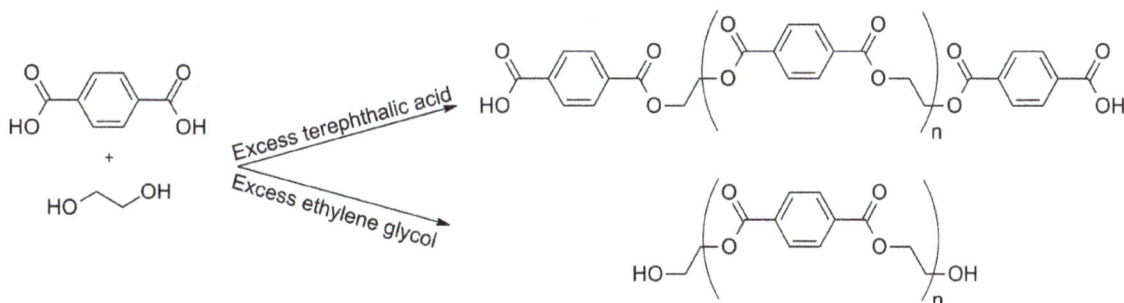

Free Radical Polymerization

The end groups that are found on polymers formed through free radical polymerization are a result from the initiators and termination method used. There are many types of initiators used in modern free radical polymerizations, and below are examples of some well-known ones. For example, azobisisobutyronitrile or AIBN forms radicals that can be used as the end groups for new starting polymer chains with styrene to form polystyrene. Once the polymer chain has formed and the reaction is terminated, the end group opposite from the initiator is a result of the terminating agent or the chain transfer agent used.

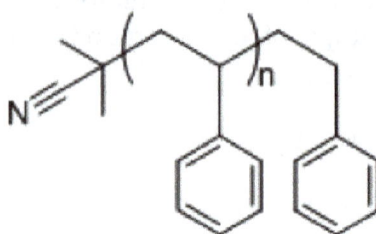

Polystyrene

Polystyrene initiated with AIBN.

Initiator	Radical Generated
Benzoyl peroxide (BPO)	
Di-*tert*-butyl peroxide (DTBP)	
Azobisisobutyronitrile (AIBN)	

End Groups in Graft Polymers

Graft copolymers are generated by attaching chains of one monomer to the main chain of another polymer; a branched block copolymer is formed. Furthermore, end groups play an important role in the process of initiation, propagation and termination of graft polymers. Graft polymers can be achieved by either "grafting from" or "grafting to"; these different methods are able to produce a vast array of different polymer structures, which can be tailored to the application in question. The "grafting from" approach involves, for example, generation of radicals along a polymer chain, which can then be reacted with monomers to grow a new polymer from the backbone of another. In "grafting from" the initiation sites on the backbone of the first polymer can be part of the backbone structure originally or generated in situ. The "grafting to" approach involves the reaction of functionalized monomers to a polymer backbone. In graft polymers, end groups play an important role, for example, in the "grafting to" technique the generation of the reactive functionalized monomers occurs at the end group, which is then tethered to the polymer chain. There are various methods to synthesize graft polymers some of the more common include redox reaction to produce free radicals, by free radical polymerization techniques avoiding chain termination (ATRP, RAFT, Nitroxide mediated, for example) and step-growth polymerization. A schematic of "grafting from" and "grafting to" is illustrated in the figure below.

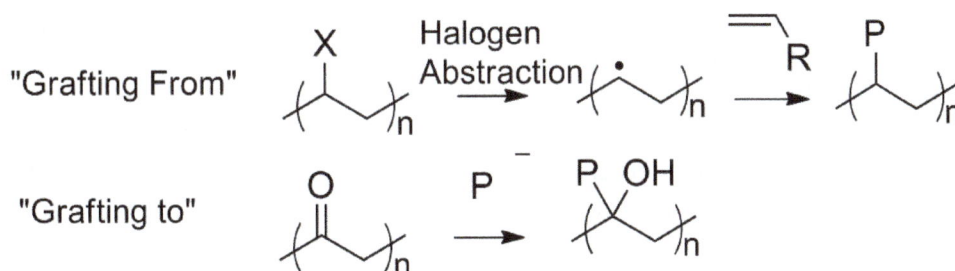

The "grafting from" technique involves the generation of radicals along the polymer backbone from an abstraction of a halogen, from either the backbone or a functional group along the backbone. Monomers are reacted with the radicals along the backbone and subsequently generate polymers which are grafted from the backbone of the first polymer. The schematic for "grafting to" shows an example using anionic polymerizations, the polymer containing the carbonyl functionalities gets attacked by the activated polymer chain and generates a polymer attached to the associated carbon along with an alcohol group, in this example. These examples show us the potential of fine tuning end groups of polymer chains to target certain copolymer structures.

Analysis of Polymers Using End Groups

Because of the importance of end groups, there have been many analytical techniques developed for the identification of the groups. The three main methods for analyzing the identity of the end group are by NMR, mass spectrometry (MS) or vibrational spectroscopy (IR or Raman). Each technique has its advantages and disadvantages, which are details below.

NMR Spectroscopy

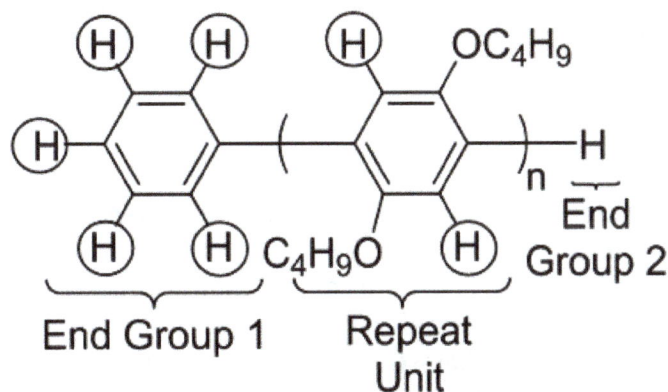

Example of utility of NMR for end group analysis.

The advantage of NMR for end groups is that it allows for not only the identification of the end group units, but also allows for the quantification of the number-average length of the polymer. End-group analysis with NMR requires that the polymer be soluble in organic or aqueous solvents. Additionally, the signal on the end-group must be visible as a distinct spectral frequency, i.e. it must not overlap with other signals. As molecular weigh increases, the width of the spectral peaks also increase. As a result of this, methods which rely on resolution of the end-group signal are mostly used for polymers of low molecular weight (roughly less than 20,000 g/mol number-average Molecular weight). By using the information obtained from the integration of a 1H NMR spectrum, the degree of polymerization (Xn) can be calculated. With knowledge of the identity of the end groups/repeat unit and the number of protons contained on each, the Xn can then be calculated. For this example above, once the 1H NMR has been integrated and the values have been normalized to 1, the degree of polymerization is calculated by simply dividing the normalized value for the repeat unit by the number of protons continued in the repeat unit. For this case, Xn = n = 100/2, and therefore Xn = 50, or there are 50 repeat units in this monomer.

Mass Spectrometry (MS)

Mass Spectrometry is helpful for the determination of the molecular weight of the polymer, structure of the polymer etc. Although chemists utilize many kinds of MS, the two that are used most typically are matrix-assisted laser desorption ionization/time of flight (MALDI-TOF) and electrospray ionization-mass spectroscopy (ESI-MS). One of the biggest disadvantages of this technique is that much like NMR spectroscopy the polymers have to be soluble in some organic solvent. An advantage of using MALDI is that it provides the simpler data to interpret for end group identification compared ESI, but a disadvantage is that the ionization can be rather hard and as a result some end groups do not remain intact for analysis. Because of the harsh ionization in MALDI, one of the biggest advantages of using ESI is for its 'softer' ionization methods. The disadvantages of using ESI is that the data obtained an be very complex due to the mechanism of the ionization and thus can be difficult to interpret.

Vibrational Spectroscopy

The vibrational spectroscopy methods used to analyze the end groups of a polymer are Infrared (IR) and Raman spectroscopy. These methods are useful in fact that the polymers do not need to be soluble in a solvent and spectra can be obtained simply from solid material. A disadvantage of the technique is that only qualitative data is typically obtained on the identification end groups.

End Group Removal

Controlled radical polymerization, namely reversible addition–fragmentation chain-transfer polymerization (RAFT), is a common method for the polymerization of acrylates, methacrylates and acrylamides. Usually, a thiocarbonate is used in combination with an effective initiator for RAFT. The thiocarbonate moiety can be functionalized at the R-group for end group analysis. The end group is a result of the propagation of chain-transfer agents during the free-radical polymerization process. The end groups can subsequently be modified by the reaction of the thiocarbonylthio compounds with nucleophiles and ionic reducing agents.

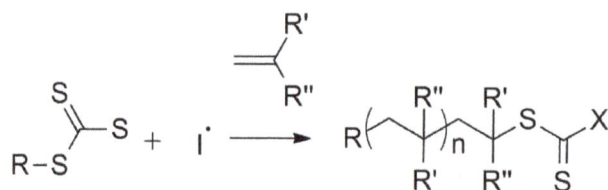

The method for removal of thiocarbonyl containing end groups includes reacting the polymers containing the end-groups with en excess of radicals which add to the reactive C=S bond of the end group forming an intermediate radical (shown below). The remaining radical on the polymer chain can be hydrogenated by what is referred to as a trapping group and terminate; this results in a polymer that is free of the end groups at the α and ω positions.

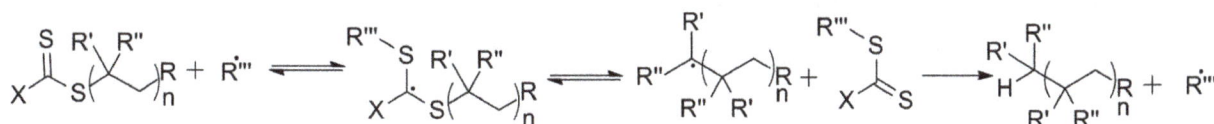

Another method of end group removal for the thiocarbonyl containing end-groups of RAFT polymers is the addition of heat to the polymer; this is referred to as thermolysis. One method of monitoring thermolysis of RAFT polymers is by thermogravietric analysis resulting in a weight-loss of the end group. An advantage of this technique is that no additional chemicals are required to remove the end group; however, it is required that the polymer be thermally stable to high temperature and therefore may not be effective for some polymers. Depending on the polymers sensitivity to ultraviolet radiation (UV) it has been reported in recent years that decomposition of end-groups can be effective, but preliminary data suggest that decomposition by UV leads to a change in the distribution of molecular weights of the polymer.

Surface Modification using RAFT

Surface modification has gained a lot of interest in recent years for a variety of applications. An example of the application of free radical polymerizations to forming new architectures is through RAFT polymerizations which result in dithioester end groups. These dithioesters can be reduced to the thiol which can be immobilized on a metal surface; this is important for applications in electronics, sensing and catalysis. The schematic below demonstrates the immobilization of copolymers onto a gold surface as reported for poly(sodium 4-styrenesulfonate) by the McCormick group at the University of Southern Mississippi.

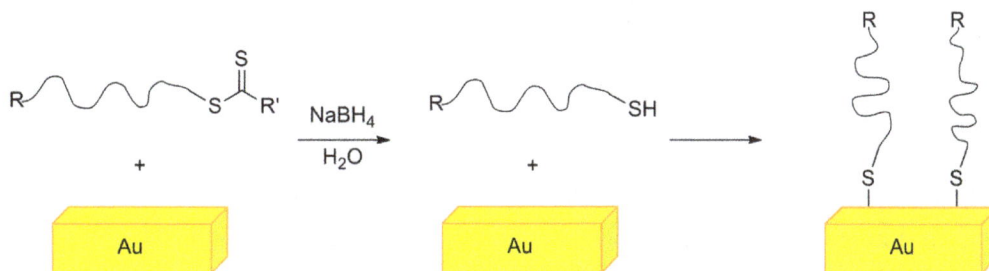

Monomer

A monomer (*mono-*, "one" + *-mer*, "part") is a molecule that may bind chemically or supramolecularly to other molecules to form a (supramolecular) polymer. The process by which monomers combine to form a polymer is called polymerization. Molecules made of a small number of monomer units (up to a few dozen) are called oligomers. The term "monomeric protein" may also be used to describe one of the proteins making up a multiprotein complex.

Industrial Polymers:

- Ethylene gas ($H_2C=CH_2$) is the precursor monomer for polyethylene

- Other modified ethylene molecules, such as tetrafluoroethylene ($F_2C=CF_2$) which leads to Teflon, vinyl chloride ($H_2C=CHCl$) which leads to PVC, styrene ($C_6H_5CH=CH_2$) which leads to polystyrene, etc.

- Epoxide monomers may be cross linked with themselves, or with the addition of a co-reactant, to form epoxy

- BPA is the monomer precursor for polycarbonate

- Many more

Biopolymer groupings, and the types of monomers that create them:

- For *lipids* (Diglycerides, triglycerides)*, the monomers are glycerol and fatty acids.

- For proteins (Polypeptides), the monomers are amino acids.

- For *Nucleic acids* (DNA/RNA), the monomers are nucleotides which is made of a pentose sugar, a nitrogenous base and a phosphate group.

- For *carbohydrates* (Polysaccharides specifically and disaccharides—depends), the monomers are monosaccharides.

*Diglycerides and triglycerides are made by dehydration synthesis from smaller molecules; this is not the same kind of end-to-end linking of similar monomers that qualifies as polymerization. Therefore, diglycerides and triglycerides are an exception to the term polymer.

Examples: The most common natural monomer is glucose, which is linked by glycosidic bonds into polymers such as cellulose, starch, and glycogen. Most often the term *monomer* refers to the organic molecules which form synthetic polymers, such as vinyl chloride, which is used to produce the polymer polyvinyl chloride (PVC).

Natural Monomers

Amino acids are natural monomers that polymerize at ribosomes to form proteins. Nucleotides, monomers found in the cell nucleus, polymerize to form nucleic acids – DNA and RNA. Glucose monomers can polymerize to form starches, glycogen or cellulose; xylose monomers can polymerise to form xylan. In all these cases and is thus not pliable, a hydrogen atom and a hydroxyl (-OH) group are lost to form H_2O, and an oxygen atom links each monomer unit. Due to the formation of water as one of the products, these reactions are known as dehydration.

Isoprene is a natural monomer and polymerizes to form natural rubber, most often *cis*-1,4-poly-isoprene, but also *trans*-1,4-polymer

Molecular Weight

The lower molecular weight compounds built from monomers are also referred to as dimers, trimers, tetramers, pentamers, hexamers, heptamers, octamers, nonamers, decamers, dodecamers, eicosamers, etc. if they have 2, 3, 4, 5, 6, 7, 8, 9, 10, 12, or 20 monomer units, respectively. Any number of these monomer units may be indicated by the appropriate Greek prefix. Larger numbers are often stated in English or numbers instead of Greek; e.g., a *20-mer* is formed from 20 monomers. Molecules made of a small number of monomer units, up to a few dozen, are called oligomers.

Industrial Use

In the light of the current tight monomers market, particularly in propylene, and of the benefits of membrane-based recovery processes, major polyolefin producers around the world already employ such recovery processes in new state-of-the-art plants. In order to enhance the competitiveness of older plants, the use of a recovery solution has started to become mandatory.

Expanding Monomers

Expanding Monomers are monomers which are increasing in volume (expanding) during polymerization. They can be added to monomer formulations to counteract the usual volumen shrinking (during polymerization) to manufacture products with higher quality and durability. Volume Shrinkage is in first line for the unmeltable thermosets a problem, since those are of fixed shape after polymerization completed.

Background

The quality of thermosets (crosslinked polymers) is determined by a numerous factors such as the purity of the used monomer, polymerization time and temperature, stoichiometry of comonomers (when used) or type and quantity of catalyst or initiator. Another rarely minded factor is the volume shrinking (and density increase) during polymerization; actually all polymers are shrinking during polymerization to some degree. This volume shrinkage can lead (after the gel point) to mechanical stress within the polymer (internal stress), which may cause microfractures, worse mechanical properties or detaching form the substrate. Expandable monomers occupy after polymerization a greater volume than before and were designed to counteract the volumetric shrinkage upon addition. For other applications, like precision castings or dental fillings, a slight expansion during polymerization would be desirable for complete filling of a given mold. Nonetheless, for some applications even a small shrinkage can be desirable as for one-piece molds, to accomplish an easy removal. Expanding monomers are used to influence respectively control the volume change during the polymerization.

Reason for Shrinkage

Shrinkage is observed during both, the polymerization and the crosslinking (curing) of monomers. This volume shrinkage is caused by various factors. The main reason is that the monomers are moving from van der Waals distance to covalent distance when a covalent bond is formed during polymerization. This can be emphasized for the example of the ethene polymerization.

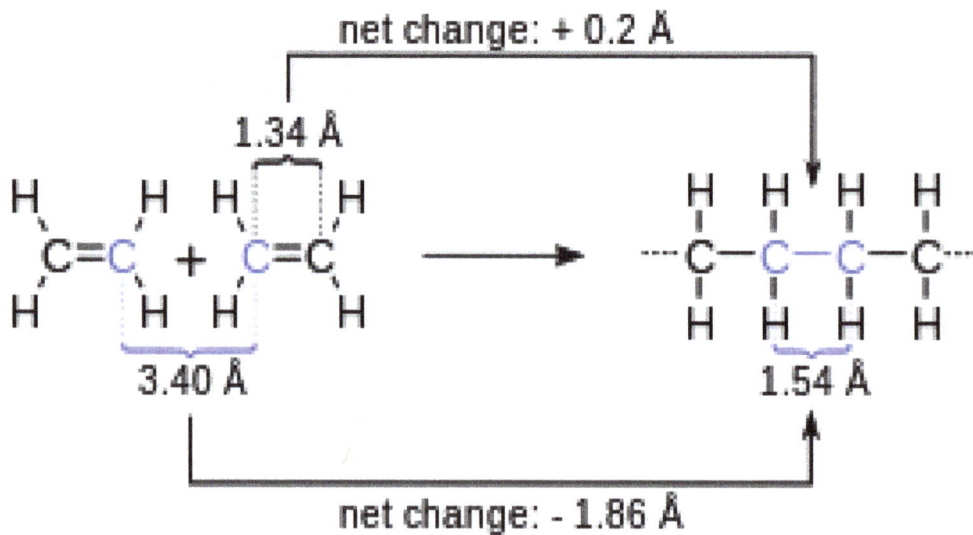

It can be seen that the distance between to monomers changes from van der Waals distance (3.40 Å) to covalent distance of a single bond (1.54 Å), leading to a net change of -1.86 Å. The change from double bond (1.34 Å) to a single bond (again 1.54 Å) results in a slight expansion (+0.2 Å). Both effects added up are still resulting in remarkable shrinkage.

A minor role is playing furthermore the entropy change while polymerization and the package density, as the polymer is more closely packed than the monomer. In step-growth polymerization (condensation reaction) small molecules are eliminated, which are also contributing to shrinkage when removed. At elevated temperatures also the thermal aging plays a role, where unreacted monomer can polymerize and degradation products and other small molecules are released.

Conventional Methods for Shrinkage Reduction

Considerable research has been done to reduce shrinkage during polymerization. As methods the addition of fillers, the use of prepolymers, the addition of reactive diluents and special crosslinking agents are conventionally used. It is a general rule that the lower the reactive portion, the lower the shrinkage of the resin is during polymerization.

Fillers (silica, mica, quartz, etc.) are reducing the shrinkage in proportion to the used amount since the volume stable filler replaces the shrinking polymer. The viscosity increase which is caused by fillers is disadvantageous since it restricts the flow of resins and mold fill. Furthermore, problematic is their tendency of settling.

Prepolymers did undergo polymerization already to some extent. However, they are still viscous and not yet gelatinised. As prepolymers are already partially polymerized the shrinking is therefore reduced during the final cure. The higher the molecular weight of the used monomers the lower the shrinkage in volume.

Also the addition of reactive diluents can reduce the shrinkage in proportion to the extent of its addition.

Concept of expanding Monomers

Ring Effects

Number of carbon atoms	Linear hydrocarbon density (g·cm⁻³)	Cyclic hydrocarbon density (g·cm⁻³)	Density difference
4	0.5788	0.6820	-0.1032
6	0.6603	0.7791	-0.1188
8	0.7028	0.8337	-0.1304
10	0.7310	0.8580	-0.1270

The expanding monomers were developed on the observation that the shrinkage during ring-opening polymerization is lower than in any other kind of polymerization:

ring-opening polymerization > chain growth polymerization > step-growth

This is mainly based on the fact that cyclic compounds are possessing higher densities than their linear counterparts. This can be illustrated by a comparison of cyclic and linear hydrocarbons (see table). A hypothetical ring-opening process of cyclobutane to n-butane would result in a volumen expansion of approximately 15%. The polymerization of a cyclic compound causes therefore a smaller volume shrinkage because cyclic compounds are already relatively dense.

It can be furthermore seen that the larger the ring, the larger the expansion. This first effect is called ring size effect. However, when the cyclic hydrocarbons would be hypothetically polymerized (to polyethelene), still a volume shrinking would appear overall (as polyethelene has a density of 0.92 g·cm-³). Nevertheless, this shrinking would be reduced with increasing ring length.

However, for a real polymerization also the ring strain has to be kept in mind. The ring strain is reduced with increasing ring size and reaches near zero in cyclohexane. This can be illustrated with the fact that oxirane polymerizes readily while oxolane is by far less reactive.

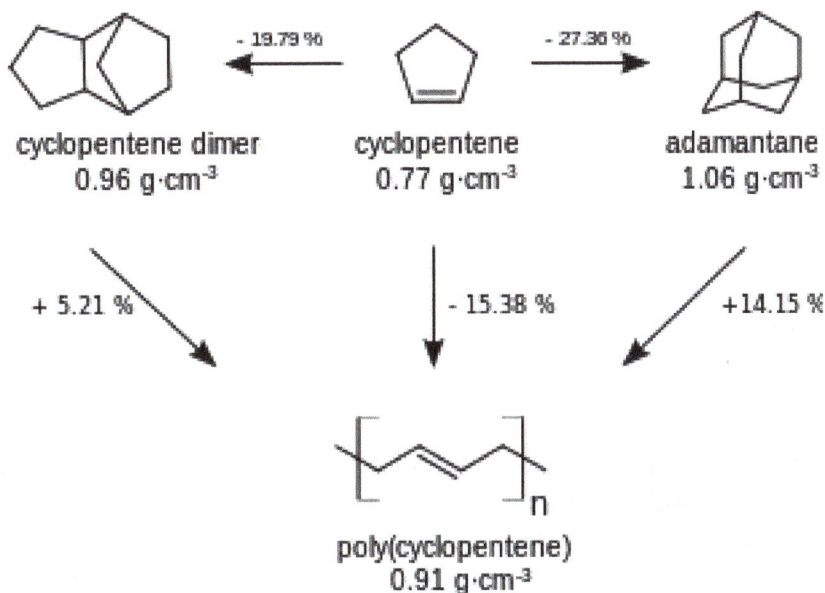

cyclopentene dimer
0.96 g·cm⁻³

cyclopentene
0.77 g·cm⁻³

adamantane
1.06 g·cm⁻³

- 19.79 %

- 27.36 %

+ 5.21 %

- 15.38 %

+14.15 %

poly(cyclopentene)
0.91 g·cm⁻³

A second effect is the ring per unit volume effect. The volume change during ring-opening polymerization is also influenced by the number of polymerizable rings per monomer. This can be illustrated for the example of cyclopentene, the cyclopentene dimer, adamantane and poly(cyclopentene). It can be seen that hypothetical conversion of cyclopentene to poly(cyclopentene) would result in volume shrinkage of 15.38%, while the conversion of cyclopentene dimer leads to expansion of 5.21% and the conversion of adamantane even to expansion of 14.15%.

The third effect is the ring-opening effect which can be illustrated at the polymerization of oxirane. During the polymerization two molecules are moving from van der Waals distance to covalent distance, what would result taken alone in shrinkage of approximately 40% (as it has been visualized above). At the same time the ring opens and moves from covalent distance to near van der Waals distance, this would result in an expansion of 17%. Thus, the overall volume change is a shrinkage of about 23%.

This minor shrinking during the ring-opening polymerization itself depends on the ring size effect, the number of rings per volume effect and the ring-opening effect.

Expanding Monomer Concept

Derived from the ring effects the design of the expanding monomers is based on bicyclic compounds. A net expansion is reached when for each bond which undergoes a shift from covalent to van der Waals distance at least two bonds are shifting from covalent to near van der Waals distance, as it is shown in the following picture.

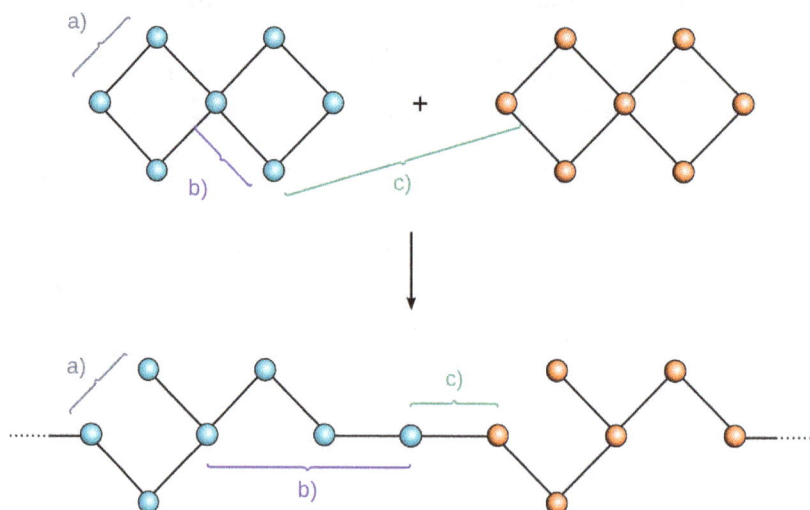

It can be seen that bond a) and bond b) are broken and changing therefore from covalent to near van der Waals distance. At the same time is bond c) is formed between two monomers, which is a change from van der waals distance to covalent distance.

It follows the three requirements that the rings of the bicyclic monomer are fused (the rings are sharing at least one atom), that each ring contains at least one non-carbon atom and that the rings are opening in asymmetrical manner (meaning for example that one oxygen forms an carbonyl group and one an ether group). Compound classes which are fulfilling these requirements are spiro orthoesters, spiro orthocarbonates, bicyclic ketal lactones and bicyclic orthoesters.

Overview, Synthesis and Polymerization

Most expanding monomers are orthoesters, either spiro orthoesters, bicyclic orthoesters or orthocarbonates. Some expanding monomers are lactones. These classes are listed in the following table.

Class of monomer	Structural formula
bicyclic orthoester	
cyclic carbonate	
spiro orthocarbonate	
spiro orthoester	
spiro orthocarbonate.	
unsaturated spiro orthoester	
bicyclic monolactone	
bicyclic bislactone	

Synthesis

There are three possibilities given in the literature for the synthesis of expanding monomers which are based on orthoesters. The first possibility is the reaction of an epoxide with an lactone:

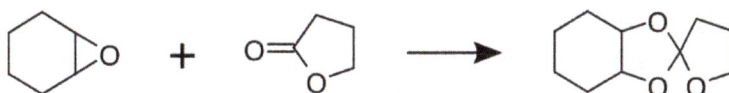

The epoxide cyclohexene oxide and the lactone γ-butyrolactone are reacting to the spiro orthoester spiro-7-9-dioxacyclo[4.3.0]nonane-8,2'1'-oxacyclo-pentane.

Also the reaction of an epoxide and an carbonate forming an spiro orthocarbonate is possible and described in literature.

The second possibility is transesterification:

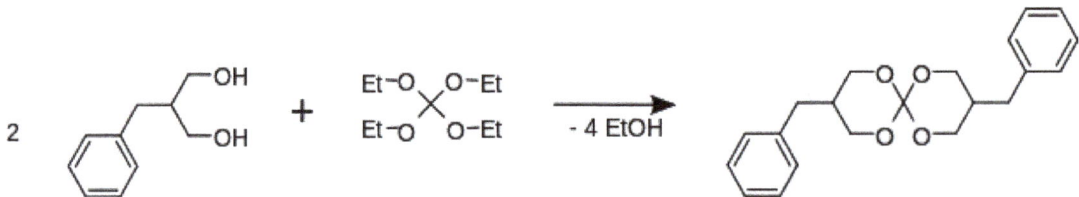

2-Benzyl-1,3-propanediol and tetraethyl orthocarbonate are reacting to 3,9-bis(phenylmethyl)-1,5,7,11-tetraoxaspiro[5.5]undecane.

Also a condensation analogous to acetalisation reaction is possible.

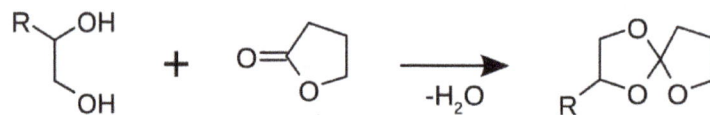

An ethanediol derivate and γ-butyrolactone are reacting to a derivate of 1,4,6-trioxa-spiro[4.4]nonane.

The third possibility is using dibutyltin oxide and carbon disulfide:

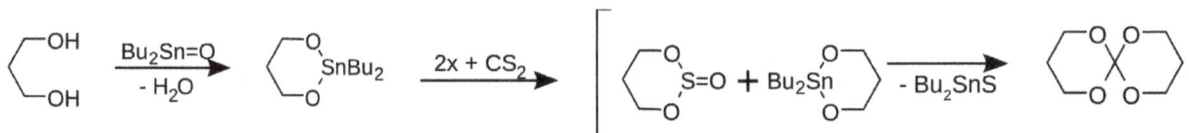

1,3-propanediol is reacting with dibutyltin oxide to 2,2-dibutyl-1,3,2-dioxastannane and carbon disulfide to the cyclic sulfite of 1,3-propanediol. Both are forming together 1,5,7,11-tetraoxa-spiro[5.5]undecane

Polymerization

Most expanding monomers are cationically polymerized, some anionically and very few even radically. Spiro orthoesters are forming, when homopolymerized, polyether polyesters.

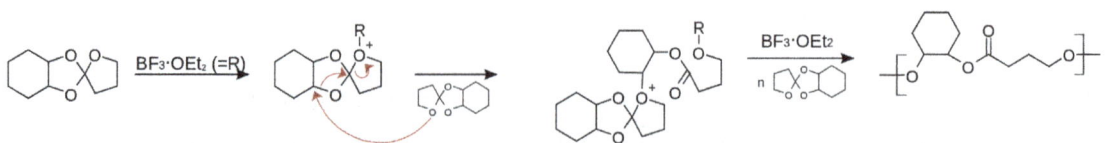

The reaction mechanism is not yet clear in details, as several side reactions are taking place. Expanding monomers can not just be homopolymerized as it is shown here but also copolymerized with other monomers to counteract their shrinking.

Usually a Lewis acid like boron trifluoride etherate is used for both, the synthesis of the orthoester and the polymerization. The same applies for spiro orthocarbonates and bicyclic orthoesters. All three are, in dependency of structure, very sensitive to moisture.

Application

Expanding monomers are interesting for application as in matrix resins in radically polymerized dental fillings, high-strength composites (e. g. in epoxy resins), adhesives, coatings, precision castings, and sealant materials to counteract shrinking during polymerization. This can be necessary in case of dental fillings since polymerization shrinkage and subsequent contraction stress in the resin composite and at the bonding interface may lead to debonding, microleakage, post-operative sensitivity, a compromise in the material's physical properties and even cracks in healthy tooth structure. They are used in the other called applications to remedy similar problems.

In recent times the UV-induced photopolymerization of spiro orthocarbonates was point of investigation.

Kuhn Length

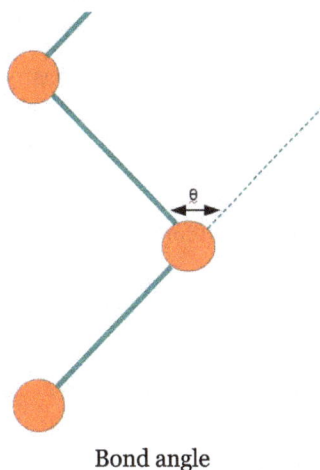

Bond angle

The Kuhn length is a theoretical treatment, developed by Werner Kuhn, in which a real polymer chain is considered as a collection of N Kuhn segments each with a Kuhn length b. Each Kuhn segment can be thought of as if they are freely jointed with each other. Each segment in a freely jointed chain can randomly orient in any direction without the influence of any forces, independent of the directions taken by other segments. Instead of considering a real chain consisting of n bonds and with fixed bond angles, torsion angles, and bond lengths, Kuhn considered an equivalent ideal chain with N connected segments, now called Kuhn segments, that can orient in any random direction.

The length of a fully stretched chain is $L = Nb$ for the Kuhn segment chain. In the simplest treatment, such a chain follows the random walk model, where each step taken in a random direction is independent of the directions taken in the previous steps, forming a random coil. The average end-to-end distance for a chain satisfying the random walk model is $\langle R^2 \rangle = Nb^2$.

Since the space occupied by a segment in the polymer chain cannot be taken by another segment, a self-avoiding random walk model can also be used. The Kuhn segment construction is useful in that it allows complicated polymers to be treated with simplified models as either a random walk or a self-avoiding walk, which can simplify the treatment considerably.

For an actual homopolymer chain (consists of the same repeat units) with bond length and bond angle θ with a dihedral angle energy potential, the average end-to-end distance can be obtained as

$$\langle R^2 \rangle = nl^2 \frac{1+\cos(\theta)}{1-\cos(\theta)} \cdot \frac{1+\langle\cos(\phi)\rangle}{1-\langle\cos(\phi)\rangle},$$

where $\langle\cos(\phi)\rangle$ is the average cosine of the dihedral angle.

The fully stretched length $L = nl\cos(\theta/2)$. By equating $\langle R^2 \rangle$ and L for the actual chain and the equivalent chain with Kuhn segments, the number of Kuhn segments N and the Kuhn segment length b can be obtained.

For worm-like chain, Kuhn length equals two times the persistence length.

Reference

- Alan D. MacNaught, Andrew R. Wilkinson, ed. (1997). Compendium of Chemical Terminology: IUPAC Recommendations (the "Gold Book") (2nd ed.). Blackwell Science. ISBN 0865426848.

- Inzelt, György (2008). "Chapter 8: Historical Background (Or: There Is Nothing New Under the Sun)". In Scholz, F. Conducting Polymers: A New Era in Electrochemistry. Monographs in Electrochemistry. Springer. pp. 265–267. ISBN 978-3-540-75929-4.

- Rosato, Dominick V.; Rosato, Donald V.; Rosato, Matthew V. (2004). Plastic product material and process selection handbook. Elsevier. p. 85. ISBN 978-1-85617-431-2.

- Mendelson, Cheryl (17 May 2005). Home Comforts: The Art and Science of Keeping House. Simon and Schuster. ISBN 9780743272865.

- Greenwood, Norman N.; Earnshaw, Alan (1997). Chemistry of the Elements (2nd ed.). Butterworth-Heinemann. p. 362. ISBN 0-08-037941-9.

- James E. Mark; Harry R. Allcock; Robert West (24 March 2005). Inorganic Polymers. Oxford University. p. 155. ISBN 978-0-19-535131-6.

- Q. Ashton Acton: Silicones—Advances in Research and Application: 2013 Edition, ScholarlyEditions, 2013, ISBN 9781481692397, p. 226.

- Wunsch, J.R. (2000). Polystyrene – Synthesis, Production and Applications. iSmithers Rapra Publishing. p. 15. ISBN 978-1-85957-191-0. Retrieved 25 July 2012.

- Campbell, Neil A.; Brad Williamson; Robin J. Heyden (2006). Biology: Exploring Life. Boston, Massachusetts: Pearson Prentice Hall. ISBN 0-13-250882-6.

- Asua, José M. (August 2007). Polymer Reaction Engineering (Hardcover - 392 pages). Wiley, John & Sons. ISBN 978-1-4051-4442-1.

- Meier, edited by John M. Chalmers, Robert J. (2008). Molecular characterization and analysis of polymers (1st ed.). Amsterdam: Elsevier. pp. 171–203. ISBN 978-0-444-53056-1.

Types of Polymers

Different polymers have different properties. Some of the most popular and common polymers manufactured are thermoplastic, polyethylene, polystyrene etc. They are manufactured keeping in mind various aspects such as use, strength and degradation. Polymers are best understood in confluence with the major topics listed in the following chapter.

Thermoplastic

A thermoplastic, or thermosoftening plastic, is a plastic material, a polymer, that becomes pliable or moldable above a specific temperature and solidifies upon cooling.

Most thermoplastics have a high molecular weight. The polymer chains associate through intermolecular forces, which weaken rapidly with increased temperature, yielding a viscous liquid. Thus, thermoplastics may be reshaped by heating and are typically used to produce parts by various polymer processing techniques such as injection molding, compression molding, calendering, and extrusion. Thermoplastics differ from thermosetting polymers, which form irreversible chemical bonds during the curing process. Thermosets do not melt, but decompose and do not reform upon cooling.

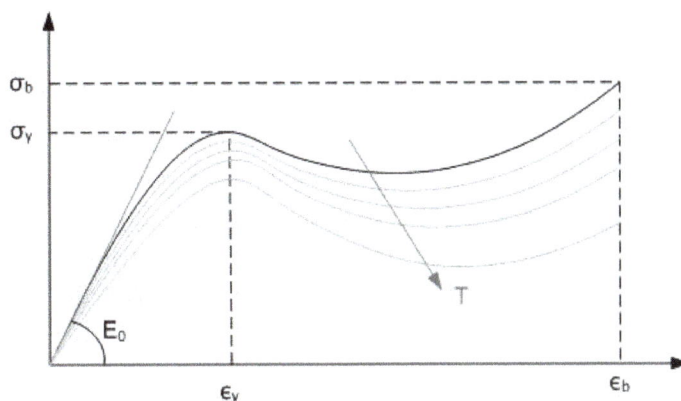

Stress-strain graph of a thermoplastic material

Above its glass transition temperature, T_g, and below its melting point, T_m, the physical properties of a thermoplastic change drastically without an associated phase change.

Some thermoplastics do not fully crystallize below the glass transition temperature T_g, retaining some or all of their amorphous characteristics. Amorphous and semi-amorphous plastics are used when high optical clarity is necessary, as light is scattered strongly by crystallites larger than its wavelength. Amorphous and semi-amorphous plastics are less resistant to chemical attack and environmental stress cracking because they lack a crystalline structure.

Brittleness can be decreased with the addition of plasticizers, which increases the mobility of amorphous chain segments to effectively lower T_g. Modification of the polymer through copolymerization or through the addition of non-reactive side chains to monomers before polymerization can also lower T_g. Before these techniques were employed, plastic automobile parts would often crack when exposed to cold temperatures.

Acrylic

Acrylic, a polymer called poly(methyl methacrylate) (PMMA), is also known by trade names such as Lucite, Perspex and Plexiglas. It serves as a sturdy substitute for glass for items such as aquariums, motorcycle helmet visors, aircraft windows, viewing ports of submersibles, and lenses of exterior lights of automobiles. It is extensively used to make signs, including lettering and logos. In medicine, it is used in bone cement and to replace eye lenses. Acrylic paint consists of PMMA particles suspended in water.

ABS

Acrylonitrile butadiene styrene (ABS) is a terpolymer synthesized from styrene and acrylonitrile in the presence of polybutadiene. ABS is a light-weight material that exhibits high impact resistance and mechanical toughness. It poses few risks to human health under normal handling. It is used in many consumer products, such as toys, appliances, and telephones.

Nylon

Nylon belongs to a class of polymers called polyamides. It has served as a substitute mainly for hemp, cotton and silk, in products such as parachutes, cords, sails, flak vests and women's clothing. Nylon fibers are useful in making fabrics, rope, carpets and musical strings, whereas in bulk form, Nylon is used for mechanical parts including machine screws, gears and power tool casings. In addition, it is used in the manufacture of heat-resistant composite materials.

PLA

Polylactic acid (polylactide) is a biodegradable thermoplastic aliphatic polyester derived from renewable resources, such as corn starch (in the United States), tapioca roots, chips or starch (mostly in Asia), or sugarcane. It is one of the materials used for 3D printing with fused deposition modeling (FDM) techniques.

Polybenzimidazole

Polybenzimidazole(PBI,short for Poly-[2,2'-(m-phenylen)-5,5'-bisbenzimidazole]) fiber is a synthetic fiber with a very high melting point. It has exceptional thermal and chemical stability and does not readily ignite. It was first discovered by American polymer chemist Carl Shipp Marvel in the pursuit of new materials with superior stability, retention of stiffness, toughness at elevated temperature. Due to its high stability, Polybenzimidazole is used to fabricate high-performance protective apparel such as firefighter's gear, astronaut space suits, high temperature protective gloves, welders' apparel and aircraft wall fabrics. In recent years, polybenzimidazole found its application as membrane in fuel cells.

Polycarbonate

Polycarbonate (PC) thermoplastics are known under trademarks such as Lexan, Makrolon, Makroclear, and arcoPlus. They are easily worked, molded, and thermoformed for many applications, such as electronic components, construction materials, data storage devices, automotive and aircraft parts, check sockets in prosthetics, and security glazing. Polycarbonates do not have a unique resin identification code. Items made from polycarbonate can contain the precursor monomer bisphenol A (BPA).

Polyether Sulfone

Polyether sulfone (PES) is a class of specially engineered thermoplastics with high thermal, oxidative, and hydrolytic stability, and good resistance to aqueous mineral acids, alkalis, salt solutions, oils and greases.

Polyetherether Ketone

Poly ether ether ketone (PEEK) has attractive properties like good abrasion resistance, low flammability and emission of smoke and toxic gases, resistance to radiation and high temperature steam, and low water absorption.

Polyetherimide

Polyetherimide (PEI), produced by a novel nitro displacement reaction involving bisphenol A, 4, 4'-methylenedianiline and 3-nitrophthalic anhydride, has high heat distortion temperature, tensile strength and modulus. They are generally used in high performance electrical and electronic parts, microwave appliances, and under-the-hood automotive parts.

Polyethylene

Polyethylene (polyethene, polythene, PE) is a family of similar materials categorized according to their density and molecular structure. For example:

- Ultra-high molecular weight polyethylene (UHMWPE) is tough and resistant to chemicals. It is used to manufacture moving machine parts, bearings, gears, artificial joints and some bulletproof vests.

- High-density polyethylene (HDPE), recyclable plastic no. 2, is commonly used as milk jugs, liquid laundry detergent bottles, outdoor furniture, margarine tubs, portable gasoline cans, drinking water distribution systems, water drainage pipes, and grocery bags.

- Medium-density polyethylene (MDPE) is used for packaging film, sacks and gas pipes and fittings.

- Low-density polyethylene (LDPE) is flexible and is used in the manufacture of squeeze bottles, milk jug caps, retail store bags and linear low-density polyethylene (LLDPE) as stretch wrap in transporting and handling boxes of durable goods, and as the common household food covering.

- XLPE or "PEX" (cross-linked polyethylene) is a semi-rigid, flexible material which has gained wide use in cold or hot water building heating and cooling applications (hydronic heating and cooling) due to its exceptional resistance to breakdown from wide temperature variations.

Polyphenylene Oxide

Polyphenylene oxide (PPO), which is obtained from the free-radical, step-growth oxidative coupling polymerization of 2,6-xylenol, has many attractive properties such as high heat distortion and impact strength, chemical stability to mineral and organic acids, and low water absorption. PPO is difficult to process, and hence the commercial resin (Noryl) is made by blending PPO with high-impact polystyrene (HIPS) which serves to reduce the processing temperature.

Polyphenylene Sulfide

Polyphenylene sulfide (PPS) obtained by the condensation polymerization of p-dichlorobenzene and sodium sulfide, has outstanding chemical resistance, good electrical properties, excellent flame retardance, low coefficient of friction and high transparency to microwave radiation. PPS is principally used in coating applications. This is done by spraying an aqueous slurry of PPS particles and heating to temperatures above 370°C. Particular grades of PPS can be used in injection and compression molding at temperatures (300 to 370°C) at which PPS particles soften and undergo apparent crosslinking. Principal applications of injection and compression molded PPS include cookware, bearings, and pump parts for service in various corrosive environments.

Polypropylene

Polypropylene (PP) is useful for such diverse products as reusable plastic food containers, microwave- and dishwasher-safe plastic containers, diaper lining, sanitary pad lining and casing, ropes, carpets, plastic moldings, piping systems, car batteries, insulation for electrical cables and filters for gases and liquids. In medicine, it is used in hernia treatment and to make heat-resistant medical equipment. Polypropylene sheets are used for stationery folders and packaging and clear storage bins. Polypropylene is defined by the recyclable plastic number 5. Although relatively inert, it is vulnerable to ultraviolet radiation and can degrade considerably in direct sunlight. Polypropylene is not as impact-resistant as the polyethylenes (HDPE, LDPE). It is also somewhat permeable to highly volatile gases and liquids.

Polystyrene

Polystyrene is manufactured in various forms that have different applications. Extruded polystyrene (PS) is used in the manufacture of disposable cutlery, CD and DVD cases, plastic models of cars and boats, and smoke detector housings. Expanded polystyrene foam (EPS) is used in making insulation and packaging materials, such as the "peanuts" and molded foam used to cushion fragile products. Extruded polystyrene foam (XPS), known by the trade name Styrofoam, is used to make architectural models and drinking cups for heated beverages. Polystyrene copolymers are used in the manufacture of toys and product casings.

Polyvinyl Chloride

Polyvinyl chloride (PVC) is a tough, lightweight material that is resistant to acids and bases. Much of it is used by the construction industry, such as for vinyl siding, drainpipes, gutters and roofing sheets. It is also converted to flexible forms with the addition of plasticizers, thereby making it useful for items such as hoses, tubing, electrical insulation, coats, jackets and upholstery. Flexible PVC is also used in inflatable products, such as water beds and pool toys.

Teflon

Teflon is a brand name of DuPont for a variety of the polymer polytetrafluoroethylene (PTFE), which belongs to a class of thermoplastics known as fluoropolymers. It is known as a coating for non-stick cookware. Being chemically inert, it is used in making containers and pipes that come in contact with reactive compounds. It is also used as a lubricant to reduce wear from friction between sliding parts, such as gears, bearings, and bushings.

Polyethylene

Polyethylene (abbreviated PE) or polyethene (IUPAC name polyethene or poly(methylene)) is the most common plastic. The annual global production is around 80 million tonnes. Its primary use is in packaging (plastic bags, plastic films, geomembranes, containers including bottles, etc.). Many kinds of polyethylene are known, with most having the chemical formula $(C_2H_4)_n$. PE is usually a mixture of similar polymers of ethylene with various values of n.

History

Polyethylene was first synthesized by the German chemist Hans von Pechmann, who prepared it by accident in 1898 while investigating diazomethane. When his colleagues Eugen Bamberger and Friedrich Tschirner characterized the white, waxy substance that he had created, they recognized that it contained long $-CH_2-$ chains and termed it *polymethylene.*

A pill box presented to a technician at ICI in 1936 made from the first pound of polyethylene

The first industrially practical polyethylene synthesis (diazomethane is a notoriously unstable substance that is generally avoided in industrial application) was discovered in 1933 by Eric Fawcett and Reginald Gibson, again by accident, at the Imperial Chemical Industries (ICI) works in Northwich, England. Upon applying extremely high pressure (several hundred atmospheres) to a mixture of ethylene and benzaldehyde they again produced a white, waxy material. Because the reaction had been initiated by trace oxygen contamination in their apparatus, the experiment was, at first, difficult to reproduce. It was not until 1935 that another ICI chemist, Michael Perrin, developed this accident into a reproducible high-pressure synthesis for polyethylene that became the basis for industrial LDPE production beginning in 1939. Because polyethylene was found to have very low-loss properties at very high frequency radio waves, commercial distribution in Britain was suspended on the outbreak of World War II, secrecy imposed and the new process was used to produce insulation for UHF and SHF coaxial cables of radar sets. During World War II, further research was done on the ICI process and in 1944 Bakelite Corporation at Sabine, Texas, and Du Pont at Charleston, West Virginia, began large-scale commercial production under license from ICI.

The breakthrough landmark in the commercial production of polyethylene began with the development of catalyst that promote the polymerization at mild temperatures and pressures. The first of these was a chromium trioxide–based catalyst discovered in 1951 by Robert Banks and J. Paul Hogan at Phillips Petroleum. In 1953 the German chemist Karl Ziegler developed a catalytic system based on titanium halides and organoaluminium compounds that worked at even milder conditions than the Phillips catalyst. The Phillips catalyst is less expensive and easier to work with, however, and both methods are heavily used industrially. By the end of the 1950s both the Phillips- and Ziegler-type catalysts were being used for HDPE production. In the 1970s, the Ziegler system was improved by the incorporation of magnesium chloride. Catalytic systems based on soluble catalysts, the metallocenes, were reported in 1976 by Walter Kaminsky and Hansjörg Sinn. The Ziegler- and metallocene-based catalysts families have proven to be very flexible at copolymerizing ethylene with other olefins and have become the basis for the wide range of polyethylene resins available today, including very low density polyethylene and linear low-density polyethylene. Such resins, in the form of UHMWPE fibers, have (as of 2005) begun to replace aramids in many high-strength applications.

Properties

The properties of polyethylene can be divided into mechanical, chemical, electrical, optical and thermal properties.

Mechanical Properties

Polyethylene is of low strength, hardness and rigidity, but has a high ductility and impact strength as well as low friction. It shows strong creep under persistent force, which can be reduced by addition of short fibers. It feels waxy when touched.

Thermal Properties

The usefulness of polyethylene is limited by its softening point of 80 °C (176 °F) (HDPE, types of low crystalline softens earlier). For common commercial grades of medium- and high-density polyethylene the melting point is typically in the range 120 to 180 °C (248 to 356 °F). The melting point for average, commercial, low-density polyethylene is typically 105 to 115 °C (221 to 239 °F). These temperatures vary strongly with the type of polyethylene.

Chemical Properties

Polyethylene consists of nonpolar, saturated, high molecular weight hydrocarbons. Therefore, its chemical behavior is similar to paraffin. The individual macromolecules are not covalently linked. Because of their symmetric molecular structure, they tend to crystallize; overall polyethylene is partially crystalline. Higher crystallinity increases density and mechanical and chemical stability.

Most LDPE, MDPE, and HDPE grades have excellent chemical resistance, meaning they are not attacked by strong acids or strong bases, and are resistant to gentle oxidants and reducing agents. Crystalline samples do not dissolve at room temperature. Polyethylene (other than cross-linked polyethylene) usually can be dissolved at elevated temperatures in aromatic hydrocarbons such as toluene or xylene, or in chlorinated solvents such as trichloroethane or trichlorobenzene.

Polyethylene absorbs almost no water. The gas and water vapor permeability (only polar gases) is lower than for most plastics; oxygen, carbon dioxide and flavorings on the other hand can pass it easily.

PE can become brittle when exposed to sunlight, carbon black is usually used as a UV stabilizer.

Polyethylene burns slowly with a blue flame having a yellow tip and gives off an odour of paraffin (similar to candle flame). The material continues burning on removal of the flame source and produces a drip.

Polyethylene can not be imprinted or stuck together without pretreatment.

Electrical Properties

Polyethylene is a good electrical insulator. It offers good tracking resistance, however, it becomes easily electrostatically charged (which can be reduced by additions of graphite, carbon black or antistatic agents).

Optical Properties

Depending on thermal history and film thickness PE can vary between almost clear (transparent), milky-opaque (translucent) or opaque. PE-LD thereby owns the largest, PE-LLD slightly lower and PE-HD the least transparency. Transparency is reduced by crystallites, if they are larger than the wavelength of visible light.

Manufacturing Process

Monomer

The ingredient or monomer is ethylene (IUPAC name ethene), a gaseous hydrocarbon with the formula C_2H_4, which can be viewed as a pair of methylene groups (=CH2) connected to each other. Because the compound is highly reactive, the ethylene must be of high purity. Typical specifications are <5 ppm for water, oxygen, and other alkenes. Acceptable contaminants include N_2, ethane (common precursor to ethylene), and methane. Ethylene is usually produced from petrochemical sources, but also is generated by dehydration of ethanol.

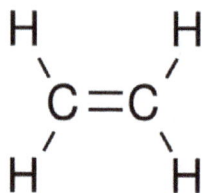

Ethylene (ethene)

Polymerization

Ethylene is a rather stable molecule that polymerizes only upon contact with catalysts. The conversion is highly exothermic. Coordination polymerization is the most pervasive technology, which means that metal chlorides or metal oxides are used. The most common catalysts consist of titanium(III) chloride, the so-called Ziegler-Natta catalysts. Another common catalyst is the Phillips catalyst, prepared by depositing chromium(VI) oxide on silica. Polyethylene can be produced through radical polymerization, but this route has only limited utility and typically requires high-pressure apparatus.

Joining

Commonly used methods for joining polyethylene parts together include:

- Hot gas welding
- Fastening
- Infrared welding
- Laser welding
- Ultrasonic welding
- Heat sealing
- Heat fusion

Adhesives and solvents are rarely used because polyethylene is nonpolar and has a high resistance to solvents. Pressure-sensitive adhesives (PSA) are feasible if the surface is flame treated or corona treated. Commonly used adhesives include:

- Dispersion of solvent-type PSAs

- Polyurethane contact adhesives

- Two-part polyurethane or epoxy adhesives

- Vinyl acetate copolymer hot melt adhesives

Classification

Polyethylene is classified by its density and branching. Its mechanical properties depend significantly on variables such as the extent and type of branching, the crystal structure, and the molecular weight. There are several types of polyethylene:

- Ultra-high-molecular-weight polyethylene (UHMWPE)

- Ultra-low-molecular-weight polyethylene (ULMWPE or PE-WAX)

- High-molecular-weight polyethylene (HMWPE)

- High-density polyethylene (HDPE)

- High-density cross-linked polyethylene (HDXLPE)

- Cross-linked polyethylene (PEX or XLPE)

- Medium-density polyethylene (MDPE)

- Linear low-density polyethylene (LLDPE)

- Low-density polyethylene (LDPE)

- Very-low-density polyethylene (VLDPE)

- Chlorinated polyethylene (CPE)

With regard to sold volumes, the most important polyethylene grades are HDPE, LLDPE, and LDPE.

Ultra-high-molecular-weight Polyethylene (UHMWPE)

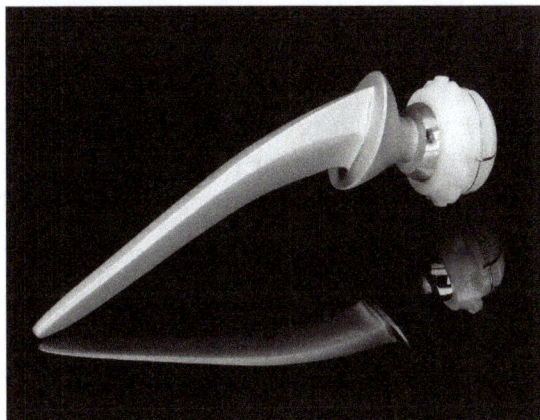

Stainless steel and ultra high molecular weight polythene hip replacement

UHMWPE is polyethylene with a molecular weight numbering in the millions, usually between 3.5 and 7.5 million. The high molecular weight makes it a very tough material, but results in less efficient packing of the chains into the crystal structure as evidenced by densities of less than high-density polyethylene (for example, 0.930–0.935 g/cm³). UHMWPE can be made through any catalyst technology, although Ziegler catalysts are most common. Because of its outstanding toughness and its cut, wear, and excellent chemical resistance, UHMWPE is used in a diverse range of applications. These include can- and bottle-handling machine parts, moving parts on weaving machines, bearings, gears, artificial joints, edge protection on ice rinks, and butchers' chopping boards. It is commonly used for the construction of articular portions of implants used for hip and knee replacements. As fiber, it competes with aramid in bulletproof vests.

High-density Polyethylene (HDPE)

HDPE pipe installation in storm drain project in Mexico

HDPE is defined by a density of greater or equal to 0.941 g/cm³. HDPE has a low degree of branching. The mostly linear molecules pack together well, so intermolecular forces are stronger than in highly branched polymers. HDPE can be produced by chromium/silica catalysts, Ziegler-Natta catalysts or metallocene catalysts; by choosing catalysts and reaction conditions, the small amount of branching that does occur can be controlled. These catalysts prefer the formation of free radicals at the ends of the growing polyethylene molecules. They cause new ethylene monomers to add to the ends of the molecules, rather than along the middle, causing the growth of a linear chain.

HDPE has high tensile strength. It is used in products and packaging such as milk jugs, detergent bottles, butter tubs, garbage containers, and water pipes. One-third of all toys are manufactured from HDPE. In 2007, the global HDPE consumption reached a volume of more than 30 million tons.

Cross-linked Polyethylene (PEX or XLPE)

PEX is a medium- to high-density polyethylene containing cross-link bonds introduced into the polymer structure, changing the thermoplastic into a thermoset. The high-temperature properties of the polymer are improved, its flow is reduced, and its chemical resistance is enhanced. PEX is used in some potable-water plumbing systems because tubes made of the material can be expanded to fit over a metal nipple and it will slowly return to its original shape, forming a permanent, water-tight connection.

Medium-density Polyethylene (MDPE)

MDPE is defined by a density range of 0.926–0.940 g/cm³. MDPE can be produced by chromium/silica catalysts, Ziegler-Natta catalysts, or metallocene catalysts. MDPE has good shock and drop resistance properties. It also is less notch-sensitive than HDPE; stress-cracking resistance is better than HDPE. MDPE is typically used in gas pipes and fittings, sacks, shrink film, packaging film, carrier bags, and screw closures.

Linear Low-density Polyethylene (LLDPE)

LLDPE is defined by a density range of 0.915–0.925 g/cm³. LLDPE is a substantially linear polymer with significant numbers of short branches, commonly made by copolymerization of ethylene with short-chain alpha-olefins (for example, 1-butene, 1-hexene, and 1-octene). LLDPE has higher tensile strength than LDPE, and it exhibits higher impact and puncture resistance than LDPE. Lower thickness (gauge) films can be blown, compared with LDPE, with better environmental stress-cracking resistance, but is not as easy to process. LLDPE is used in packaging, particularly film for bags and sheets. Lower thickness may be used compared to LDPE. It is used for cable coverings, toys, lids, buckets, containers, and pipe. While other applications are available, LLDPE is used predominantly in film applications due to its toughness, flexibility, and relative transparency. Product examples range from agricultural films, Saran wrap, and bubble wrap, to multilayer and composite films. In 2013, the world LLDPE market reached a volume of US$40 billion.

Low-density Polyethylene (LDPE)

LDPE is defined by a density range of 0.910–0.940 g/cm³. LDPE has a high degree of short- and long-chain branching, which means that the chains do not pack into the crystal structure as well. It has, therefore, less strong intermolecular forces as the instantaneous-dipole induced-dipole attraction is less. This results in a lower tensile strength and increased ductility. LDPE is created by free-radical polymerization. The high degree of branching with long chains gives molten LDPE unique and desirable flow properties. LDPE is used for both rigid containers and plastic film applications such as plastic bags and film wrap. In 2013, the global LDPE market had a volume of almost US$33 billion.

The radical polymerization process used to make LDPE does not include a catalyst that "supervises" the radical sites on the growing PE chains. (In HDPE synthesis, the radical sites are at the ends of the PE chains, because the catalyst stabilizes their formation at the ends.) Secondary radicals (in the middle of a chain) are more stable than primary radicals (at the end of the chain), and tertiary radicals (at a branch point) are more stable yet. Each time an ethylene monomer is added, it creates a primary radical, but often these will rearrange to form more stable secondary or tertiary radicals. Addition of ethylene monomers to the secondary or tertiary sites creates branching.

Very-low-density Polyethylene (VLDPE)

VLDPE is defined by a density range of 0.880–0.915 g/cm³. VLDPE is a substantially linear polymer with high levels of short-chain branches, commonly made by copolymerization of ethylene with short-chain alpha-olefins (for example, 1-butene, 1-hexene and 1-octene). VLDPE is most commonly produced using metallocene catalysts due to the greater co-monomer incorporation exhibited by these catalysts. VLDPEs are used for hose and tubing, ice and frozen food bags, food packaging and stretch wrap as well as impact modifiers when blended with other polymers.

Recently, much research activity has focused on the nature and distribution of long chain branches in polyethylene. In HDPE, a relatively small number of these branches, perhaps one in 100 or 1,000 branches per backbone carbon, can significantly affect the rheological properties of the polymer.

Copolymers

In addition to copolymerization with alpha-olefins, ethylene can also be copolymerized with a wide range of other monomers and ionic composition that creates ionized free radicals. Common examples include vinyl acetate (the resulting product is ethylene-vinyl acetate copolymer, or EVA, widely used in athletic-shoe sole foams) and a variety of acrylates. Applications of acrylic copolymer include packaging and sporting goods, and superplasticizer, used for cement production.

Molecular Structure of Different PE Types

The diverse material behavior of different types of polyethylene can be explained by their molecular structure. Molecular weight and crystallinity are having the biggest impact, the crystallinity in turn depends on molecular weight and degree of branching. The less the polymer chains are branched, and the smaller the molecular weight, the higher the crystallinity of polyethylene. The crystallinity is between 35% (PE-LD/PE-LLD) and 80% (PE-HD). Within crystallites polyethylene has a density of 1.0 g·cm^{-3}, in the amorphous regions of 0.86 g·cm^{-3}. Thus, an almost linear relationship exists between density and crystallinity.

The degree of branching of the different types of polyethylene can be schematically represented as follows:

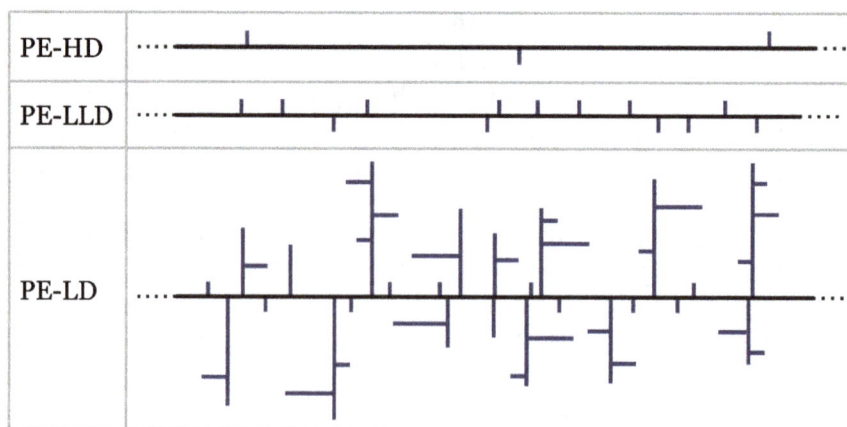

The figure shows polyethylene backbones, short-chain branches and side chain branches. The polymer chains are represented linearly.

Chain Branches

The properties of polyethylene are highly dependent on type and number of chain branches. The chain branches in turn depend on the process used: either the high-pressure process (only PE-LD) or the low-pressure process (all other PE grades). Low-density polyethylene is produced by the high-pressure process by radical polymerization, thereby numerous short chain branches as well as long chain branches are formed. Short chain branches are formed by intramolecular chain transfer reactions, they are always butyl or ethyl chain branches because the reaction proceeds after the following mechanism:

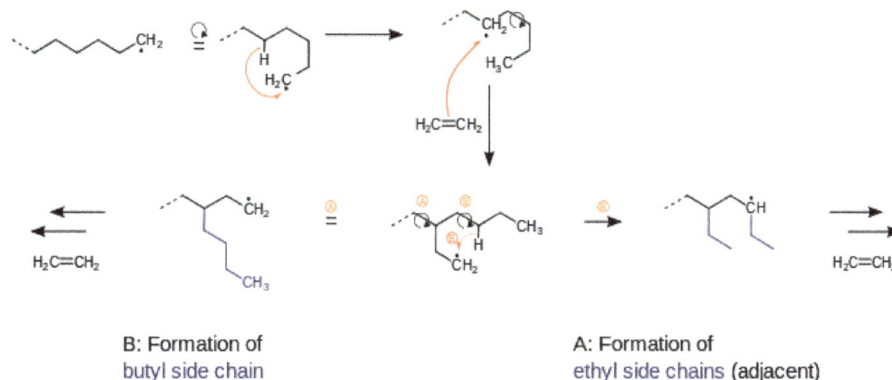

B: Formation of
butyl side chain

A: Formation of
ethyl side chains (adjacent)

Environmental Issues

A bag manufactured from polyethylene

Although ethylene can be produced from renewables, polyethylene is mainly made from petroleum or natural gas.

Biodegrading Plastics

One of the main problems of polyethylene is that without special treatment it is not readily biodegradable, and thus accumulates. In Japan, getting rid of plastics in an environmentally friendly way was the major problem discussed until the Fukushima disaster in 2011. It was listed as a $90 billion market for solutions. Since 2008, Japan has rapidly increased the recycling of plastics, but still has a large amount of plastic wrapping which goes to waste.

In May 2008, Daniel Burd, a 16-year-old Canadian, won the Canada-Wide Science Fair in Ottawa after discovering that *Pseudomonas fluorescens*, with the help of *Sphingomonas*, can degrade over 40% of the weight of plastic bags in less than three months.

The thermophilic bacterium *Brevibacillus borstelensis* (strain 707) was isolated from a soil sample and found to use low-density polyethylene as a sole carbon source when incubated together at 50 °C. Biodegradation increased with time exposed to ultraviolet radiation.

In 2010, a Japanese researcher, Akinori Ito, released the prototype of a machine which creates oil from polyethylene using a small, self-contained vapor distillation process.

Acinetobacter sp. 351 can degrade lower molecular-weight PE oligomers. When PE is subjected to thermo- and photo-oxidization, products including alkanes, alkenes, ketones, aldehydes, alcohols, carboxylic acid, keto-acids, dicarboxylic acids, lactones, and esters are released.

In 2014, a Chinese researcher discovered that Indian mealmoth larvae could metabolize polyethylene from observing that plastic bags at his home had small holes in them. Deducing that the hungry larvae must have digested the plastic somehow, he and his team analyzed their gut bacteria and found a few that could use plastic as their only carbon source. Not only could the bacteria from the guts of the *Plodia interpunctella* moth larvae metabolize polyethylene, they degraded it significantly, dropping its tensile strength by 50%, its mass by 10% and the molecular weights of its polymeric chains by 13%.

Chemically Modified Polyethylene

Polyethylene may either be modified in the polymerization by polar or non-polar comonomers or after polymerization through polymer-analogous reactions. Common polymer-analogous reactions are in case of polyethylene crosslinking, chlorination and sulfochlorination.

Non-polar Ethylene Copolymers

α-olefins

In the low pressure process α-olefins (e.g. 1-butene or 1-hexene) may be added, which are incorporated in the polymer chain during polymerization. These copolymers introduce short side chains, thus crystallinity and density are reduced. As explained above, mechanical and thermal properties are changed thereby. In particular, PE-LLD is produced this way.

Metallocene Polyethylene (PE-MC)

Metallocene polyethylene (PE-M) is prepared by means of metallocene catalysts, usually including copolymers (z. B. ethene / hexene). Metallocene polyethylene has a relatively narrow molecular weight distribution, exceptionally high toughness, excellent optical properties and a uniform comonomer content. Because of the narrow molecular weight distribution it behaves less pseudoplastic (especially under larger shear rates). Metallocene polyethylene has a low proportion of low molecular weight (extractable) components and a low welding and sealing temperature. Thus, it is particularly suitable for the food industry.

Polyethylene with Multimodal Molecular Weight Distribution

Polyethylene with multimodal molecular weight distribution consists of several polymer fractions, which are homogeneously mixed. Such polyethylene types offer extremely high stiffness, toughness, strength, stress crack resistance and an increased crack propagation resistance. They consist of equal proportions higher and lower molekularerer polymer fractions. The lower molecular weight units crystallize easier and relax faster. The higher molecular weight fractions form linking molecules between crystallites, thereby increasing toughness and stress crack resistance. Polyethylene with multimodal molecular weight distribution can be prepared either in two-stage reactors, by catalysts with two different active centers on a carrier or by blending in extruders.

Cyclic Olefin Copolymers (COC)

Cyclic olefin copolymers are prepared by copolymerization of ethene and cycloolefins (usually norbornene) produced by using metallocene catalysts. The resulting polymers are amorphous polymers and particularly transparent and heat resistant.

Polar Ethylene Copolymers

The basic compounds used as polar comonomers are vinyl alcohol (Ethenol, an unsaturated alcohol), acrylic acid (propenoic acid, an unsaturated acid) and esters containing one of the two compounds.

Ethylene Copolymers with Unsaturated Alcohols

Ethylene/vinyl alcohol copolymer (EVOH) is (formally) a copolymer of PE and vinyl alcohol (ethenol), which is prepared by (partial) hydrolysis of ethylene-vinyl acetate copolymer (as vinyl alcohol itself is not stable). However, typically EVOH has a higher comonomer content than the VAC commonly used.

EVOH is used in multilayer films for packaging as a barrier layer (barrier plastic). As EVOH is hygroscopic (water-attracting), it absorbs water from the environment, whereby it loses its barrier effect. Therefore, it must be used as a core layer surrounded by other plastics (like LDPE, PP, PA or PET). EVOH is also used as a coating agent against corrosion at street lights, traffic light poles and noise protection walls.

Ethylene/Acrylic Acid Copolymers (EAA)

Copolymer of ethylene and unsaturated carboxylic acids (such as acrylic acid) are characterized by good adhesion to different materials, by resistance to stress cracking and high flexibility. However, they are more sensitive to heat and oxidation than ethylene homopolymers. Ethylene/acrylic acid copolymers are used as adhesion promoters.

If salts of an unsaturated carboxylic acid are present in the polymer, thermo-reversible ion networks are formed, they are called ionomers. Ionomers are highly transparent thermoplastics which are characterized by high adhesion to metals, high abrasion resistance and high water absorption.

Ethylene Copolymers with Unsaturated Esters

If unsaturated esters are copolymerized with ethylene, either the alcohol moiety may be in the polymer backbone (as it is the case in ethylene-vinyl acetate copolymer) or of the acid moiety (e. g. in ethylene-ethyl acrylate copolymer). Ethylene-vinyl acetate copolymers are prepared similarly to LD-PE by high pressure polymerization. The proportion of comonomer has a decisive influence on the behaviour of the polymer.

The density decreases up to a comonomer share of 10% because of the disturbed crystal formation. With higher proportions it approaches to the one of polyvinyl acetate (1.17 g/cm^3). Due to decreasing crystallinity ethylene vinyl acetate copolymers are getting softer with increasing comonomer content. The polar side groups change the chemical properties significantly (compared to polyethylene): weather resistance, adhesiveness and weldability rise with comonomer content, while the chemical resistance decreases. Also mechanical properties are changed: stress cracking resistance and toughness in the cold rise, whereas yield stress and heat resistance decrease. With a very high proportion of comonomers (about 50%) rubbery thermoplastics are produced (thermoplastic elastomers).

Ethylene-ethyl acrylate copolymers behave similarly to ethylene-vinyl acetate copolymers.

Crosslinking

Various methods can used to prepare cross-linked polyethylene (PE-X) from thermoplastic polyethylene (PE-LD, PE-LLD or PE-HD). By crosslinking low-temperature impact strength, abrasion resistance and environmental stress cracking resistance can be increased significantly, whereas hardness and rigidity are somewhat reduced. PE-X does not melt anymore (analogous to elastomers) and is thermally resistant (over longer periods of up to 120 °C, for short periods without mechanical load up to 250 °C). With increasing crosslinking density also the maximum shear modulus increases (even at higher temperatures). PE-X has significantly enhanced properties compared with ordinary PE. As PE-X is infusible, always the final pre-products or the mold part are cross-linked.

Applications

PE-X is used as insulating material for medium and high voltage cable insulation, for hot water pipes and molded parts in electrical engineering, plant engineering and in automotive industry.

Types of Crosslinking

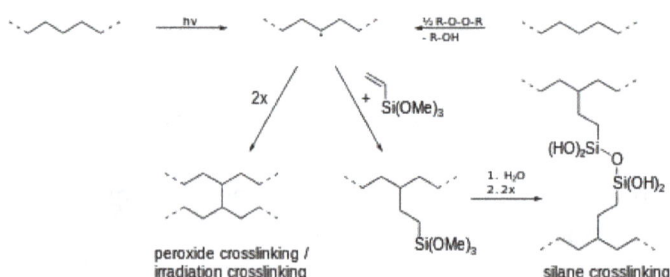

Shown are the peroxide, the silane and irradiation crosslinking. In each method, a radical is generated in the polyethylene chain (top center), either by radiation (h·v) or by peroxides (R-O-O-R). Then, two radical chains can either directly crosslink (bottom left) or indirectly by silane compounds (bottom right).

A basic distinction is made between peroxide crosslinking (PE-Xa), silane crosslinking (PE-Xb), electron beam crosslinking (PE-Xc) and azo crosslinking (PE-Xd).

- Peroxide crosslinking (PE-Xa): The crosslinking of polyethylene using peroxides (e. g. dicumyl or di-tert-butyl peroxide) is still of major importance. In the so-called *Engel process*, a mixture of HDPE and 2 % peroxide is at first mixed at low temperatures in an extruder and then crosslinked at high temperatures (between 200 and 250 °C). The peroxide decomposes to peroxide radicals (RO•), which abstract (remove) hydrogen atoms from the polymer chain, leading to radicals. When these combine, a crosslinked network is formed. The resulting polymer network is uniform, of low tension and high flexibility, whereby it is softer and tougher than (the irradiated) PE-Xc.

- Silane crosslinking (PE-Xb): In the presence of silanes (e.g. trimethoxyvinylsilane) polyethylene can initially be Si-functionalized by irradiation or by a small amount of a peroxide. Later Si-OH groups can be formed in a water bath by hydrolysis, which condense then and crosslink the PE by the formation of Si-O-Si bridges. Catalysts such as dibutyltin dilaurate may accelerate the reaction.

- Irradiation crosslinking (PE-Xc): The crosslinking of polyethylene is also possible by a downstream radiation source (usually a electron accelerator, occasionally a isotopic radiator). PE products are crosslinked below the crystalline melting point by splitting off hydrogen atoms. β-radiation possesses a penetration depth of 10 mm, γ-radiation 100 mm. Thereby the interior or specific areas can be excluded from the crosslinking. However, due to high capital and operating costs radiation crosslinking plays only a minor role compared with the peroxide crosslinking. In contrast to peroxide crosslinking, the process is carried out in the solid state. Thereby, the cross-linking takes place primarily in the amorphous regions, while the crystallinity remains largely intact.

- Azo crosslinking (PE-Xd): In the so-called *Lubonyl process* polyethylene is crosslinked preadded azo compounds after extrusion in a hot salt bath.

Degree of Crosslinking

A low degree of crosslinking leads initially only to a multiplication of the molecular weight. The individual macromolecules are not linked and no covalent network is formed yet. Polyethylene that consists of those large molecules behaves similar to polyethylene of ultra high molecular weight (PE-UHMW), i.e. like a thermoplastic elastomer.

Upon further crosslinking (crosslinking degree about 80%), the individual macromolecules are eventually connected to a network. This crosslinked polyethylene (PE-X) is chemically seen a thermoset, it shows above the melting point rubber-elastic behavior and can not be processed in the melt anymore.

The degree of crosslinking (and hence the extent of the change) is different in intensity depending on the process. According to DIN 16892 (quality requirement for pipes made of PE-X) at least the following degree of crosslinking must be achieved:

- in peroxide crosslinking (PE-Xa): 75%

- with silane crosslinking (PE-Xb): 65%

- with electron beam crosslinking (PE-Xc): 60%

- in azo crosslinking (PE-Xd): 60%

Chlorination and Sulfochlorination

Chlorinated Polyethylene (PE-C) is an inexpensive material having a chlorine content from 34 to 44%. It is used in blends with PVC because the soft, rubbery chloropolyethylene is embedded in the PVC matrix, thereby increasing the impact resistance. In addition, it also increases the weather resistance. Furthermore, it is used for softening PVC foils, without risking the migrate of plasticizers. Chlorinated polyethylene can be crosslinked peroxidically to form an elastomer which is used in cable and rubber industry. When chlorinated polyethylene is added to other polyolefins, it reduces the flammability.

Chlorosulfonated PE (CSM) is used as starting material for ozone resistant synthetic rubber.

Bio-based Polyethylene

Braskem and Toyota Tsusho Corporation started joint marketing activities to produce polyethylene from sugarcane. Braskem will build a new facility at their existing industrial unit in Triunfo, RS, Brazil with an annual production capacity of 200,000 short tons (180,000,000 kg), and will produce high-density and low-density polyethylene from bioethanol derived from sugarcane.

Polyethylene can also be made from other feedstocks, including wheat grain and sugar beet. These developments are using renewable resources rather than fossil fuel, although the issue of plastic source is currently negligible in the wake of plastic waste and in particular polyethylene waste as shown above.

Nomenclature and General Description of the Process

The name polyethylene comes from the ingredient and not the resulting chemical compound, which contains no double bonds. The scientific name *polyethene* is systematically derived from the scientific name of the monomer. The alkene monomer converts to a long, sometimes *very* long, alkane in the polymerization process. In certain circumstances it is useful to use a structure-based nomenclature; in such cases IUPAC recommends poly(methylene) (poly(methanediyl) is a non-preferred alternative). The difference in names between the two systems is due to the *opening up* of the monomer's double bond upon polymerization. The name is abbreviated to PE. In a similar manner polypropylene and polystyrene are shortened to PP and PS, respectively. In the United Kingdom the polymer is commonly called polythene, from the ICI trade name, although this is not recognized scientifically.

Polytetrafluoroethylene

Polytetrafluoroethylene (PTFE) is a synthetic fluoropolymer of tetrafluoroethylene that has numerous applications. The best known brand name of PTFE-based formulas is Teflon by Chemours. Chemours is a spin-off of DuPont Co., which discovered the compound in 1938.

PTFE is a fluorocarbon solid, as it is a high-molecular-weight compound consisting wholly of carbon and fluorine. PTFE is hydrophobic: neither water nor water-containing substances wet PTFE, as fluorocarbons demonstrate mitigated London dispersion forces due to the high electronegativity of fluorine. PTFE has one of the lowest coefficients of friction of any solid.

PTFE is used as a non-stick coating for pans and other cookware. It is very non-reactive, partly because of the strength of carbon–fluorine bonds, and so it is often used in containers and pipework for reactive and corrosive chemicals. Where used as a lubricant, PTFE reduces friction, wear and energy consumption of machinery. It is commonly used as a graft material in surgical interventions. Also, it is frequently employed as coating on catheters; this interferes with the ability of bacteria and other infectious agents to adhere to catheters and cause hospital-acquired infections.

History

Teflon thermal cover showing impact craters, from NASA's Ultra Heavy Cosmic Ray Experiment (UHCRE)

PTFE was accidentally discovered in 1938 by Roy Plunkett while he was working in New Jersey for DuPont. As Plunkett attempted to make a new chlorofluorocarbon refrigerant, the tetrafluoroethylene gas in its pressure bottle stopped flowing before the bottle's weight had dropped to the point signaling "empty." Since Plunkett was measuring the amount of gas used by weighing the bottle, he became curious as to the source of the weight, and finally resorted to sawing the bottle apart. He found the bottle's interior coated with a waxy white material that was oddly slippery. Analysis showed that it was polymerized perfluoroethylene, with the iron from the inside of the container having acted as a catalyst at high pressure. Kinetic Chemicals patented the new fluorinated plastic (analogous to the already known polyethylene) in 1941, and registered the Teflon trademark in 1945.

By 1948, DuPont, which founded Kinetic Chemicals in partnership with General Motors, was producing over two million pounds (900 tons) of Teflon brand PTFE per year in Parkersburg, West Virginia. An early use was in the Manhattan Project as a material to coat valves and seals in the pipes holding highly reactive uranium hexafluoride at the vast K-25 uranium enrichment plant in Oak Ridge, Tennessee.

In 1954, the wife of French engineer Marc Grégoire urged him to try the material he had been using on fishing tackle on her cooking pans. He subsequently created the first Teflon-coated, non-stick pans under the brandname Tefal (combining "Tef" from "Teflon" and "al" from aluminum). In the United States, Marion A. Trozzolo, who had been using the substance on scientific utensils, marketed the first US-made Teflon-coated pan, "The Happy Pan", in 1961.

However, Tefal was not the only company to utilize PTFE in nonstick cookware coatings. In subsequent years, many cookware manufacturers developed proprietary PTFE-based formulas, including Swiss Diamond International, which uses a diamond-reinforced PTFE formula; Scanpan, which uses a titanium-reinforced PTFE formula; and both All-Clad and Newell Rubbermaid's Calphalon, which use a non-reinforced PTFE-based nonstick. Other cookware companies, such as Meyer Corporation's Anolon, use Teflon nonstick coatings purchased from DuPont.

In the 1990s, it was found that PTFE could be radiation cross-linked above its melting point in an oxygen-free environment. Electron beam processing is one example of radiation processing. Cross-linked PTFE has improved high-temperature mechanical properties and radiation stability. This was significant because, for many years, irradiation at ambient conditions had been used to break down PTFE for recycling. The radiation-induced chain scission allows it to be more easily reground and reused.

Production

PTFE is produced by free-radical polymerization of tetrafluoroethylene. The net equation is

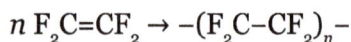

$$n \; F_2C{=}CF_2 \rightarrow -(F_2C{-}CF_2)_n-$$

Because tetrafluoroethylene can explosively decompose to tetrafluoromethane and carbon, special apparatus is required for the polymerization to prevent hot spots that might initiate this dangerous side reaction. The process is typically initiated with persulfate, which homolyzes to generate sulfate radicals:

$$[O_3SO{-}OSO_3]^{2-} \rightleftharpoons 2 \; SO_4{\bullet}^-$$

The resulting polymer is terminated with sulfate ester groups, which can be hydrolyzed to give OH end-groups.

Because PTFE is poorly soluble in almost all solvents, the polymerization is conducted as an emulsion in water. This process gives a suspension of polymer particles. Alternatively, the polymerization is conducted using a surfactant such as PFOS.

Properties

PTFE is often used to coat non-stick pans as it is hydrophobic and possesses fairly high heat resistance.

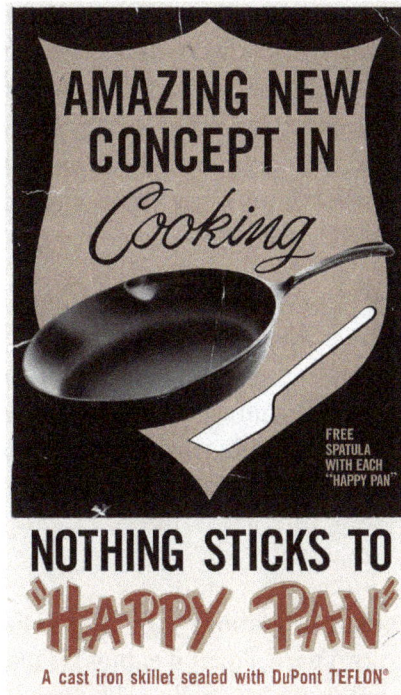

"Amazing New Concept in Cooking"

PTFE is a thermoplastic polymer, which is a white solid at room temperature, with a density of about 2200 kg/m³. According to DuPont, its melting point is 600 K (327 °C; 620 °F). It maintains high strength, toughness and self-lubrication at low temperatures down to 5 K (−268.15 °C; −450.67 °F), and good flexibility at temperatures above 194 K (−79 °C; −110 °F). PTFE gains its properties from the aggregate effect of carbon-fluorine bonds, as do all fluorocarbons. The only chemicals known to affect these carbon-fluorine bonds are highly reactive metals like the alkali metals, and at higher temperatures also such metals as aluminium and magnesium, and fluorinating agents such as xenon difluoride and cobalt(III) fluoride.

Property	Value
Density	2200 kg/m³
Glass temperature	388 K
Melting point	600 K
Thermal expansion	$112\text{--}125 \cdot 10^{-6}$ K^{-1}
Thermal diffusivity	0.124 mm²/s
Young's modulus	0.5 GPa
Yield strength	23 MPa
Bulk resistivity	10^{16} Ω·m
Coefficient of friction	0.05–0.10
Dielectric constant	$\varepsilon = 2.1$, tan(δ) < 5(-4)
Dielectric constant (60 Hz)	$\varepsilon = 2.1$, tan(δ) < 2(-4)
Dielectric strength (1 MHz)	60 MV/m

The coefficient of friction of plastics is usually measured against polished steel. PTFE's coefficient of friction is 0.05 to 0.10, which is the third-lowest of any known solid material (BAM being the first, with a coefficient of friction of 0.02; diamond-like carbon being second-lowest at 0.05). PTFE's resistance to van der Waals forces means that it is the only known surface to which a gecko cannot stick. In fact, PTFE can be used to prevent insects climbing up surfaces painted with the material. PTFE is so slippery that insects cannot get a grip and tend to fall off. For example, PTFE is used to prevent ants climbing out of formicaria.

Because of its chemical inertness, PTFE cannot be cross-linked like an elastomer. Therefore, it has no "memory" and is subject to creep. Because of its superior chemical and thermal properties, PTFE is often used as a gasket material. However, because of the propensity to creep, the long-term performance of such seals is worse than for elastomers which exhibit zero, or near-zero, levels of creep. In critical applications, Belleville washers are often used to apply continuous force to PTFE gaskets, ensuring a minimal loss of performance over the lifetime of the gasket.

Applications and Uses

The major application of PTFE, consuming about 50% of production, is for wiring in aerospace and computer applications (e.g. hookup wire, coaxial cables). This application exploits the fact that PTFE has excellent dielectric properties. This is especially true at high radio frequencies, making it suitable for use as an insulator in cables and connector assemblies and as a material for printed circuit boards used at microwave frequencies. Combined with its high melting temperature, this makes it the material of choice as a high-performance substitute for the weaker and lower-melting-point polyethylene commonly used in low-cost applications.

In industrial applications, owing to its low friction, PTFE is used for applications where sliding action of parts is needed: plain bearings, gears, slide plates, etc. In these applications, it performs significantly better than nylon and acetal; it is comparable to ultra-high-molecular-weight polyethylene (UHMWPE). Although UHMWPE is more resistant to wear than PTFE, for these applications, versions of PTFE with mineral oil or molybdenum disulfide embedded as additional lubricants in its matrix are being manufactured. Its extremely high bulk resistivity makes it an ideal material for fabricating long-life electrets, useful devices that are the electrostatic analogues of magnets.

PTFE film is also widely used in the production of carbon fiber composites as well as fiberglass composites, notably in the aerospace industry. PTFE film is used as a barrier between the carbon or fiberglass part being built, and breather and bagging materials used to incapsulate the bondment when debulking (vacuum removal of air from between layers of laid-up plies of material) and when curing the composite, usually in an autoclave. The PTFE, used here as a film, prevents the non-production materials from sticking to the part being built, which is sticky due to the carbon-graphite or fiberglass plies being pre-pregnated with bismaleimide resin. Non-production materials such as Teflon, Airweave Breather and the bag itself would be considered F.O.D. (foreign object debris/damage) if left in layup.

Because of its extreme non-reactivity and high temperature rating, PTFE is often used as the liner in hose assemblies, expansion joints, and in industrial pipe lines, particularly in applications using acids, alkalis, or other chemicals. Its frictionless qualities allow improved flow of highly viscous liquids, and for uses in applications such as brake hoses.

Gore-Tex is a material incorporating a fluoropolymer membrane with micropores. The roof of the Hubert H. Humphrey Metrodome in Minneapolis, US, was one of the largest applications of PTFE coatings. 20 acres (81,000 m²) of the material was used in the creation of the white double-layered PTFE-coated fiberglass dome.

Other

PTFE (Teflon) is best known for its use in coating non-stick frying pans and other cookware, as it is hydrophobic and possesses fairly high heat resistance.

PTFE tapes with pressure-sensitive adhesive backing

Some iron sole plate (of clothes iron) are in PTFE (Teflon).

Niche

PTFE is a versatile material that is found in many niche applications:

- It can be stretched to contain small pores of varying sizes and is then placed between fabric layers to make a waterproof, breathable fabric in outdoor apparel.

- It is used widely as a fabric protector to repel stains on formal school-wear, like uniform blazers.

- It is used as a film interface patch for sports and medical applications, featuring a pressure-sensitive adhesive backing, which is installed in strategic high friction areas of footwear, insoles, ankle-foot orthosis, and other medical devices to prevent and relieve friction-induced blisters, calluses and foot ulceration.

- Expanded PTFE membranes have been used in trials to assist trabeculectomy surgery to treat glaucoma.

- Powdered PTFE is used in pyrotechnic compositions as an oxidizer with powdered metals such as aluminium and magnesium. Upon ignition, these mixtures form carbonaceous soot and the corresponding metal fluoride, and release large amounts of heat. They are used in infrared decoy flares and as igniters for solid-fuel rocket propellants. Aluminium and PTFE is also used in some thermobaric fuel compositions.

- In optical radiometry, sheets of PTFE are used as measuring heads in spectroradiometers and broadband radiometers (e.g., illuminance meters and UV radiometers) due to PTFE's capability to diffuse a transmitting light nearly perfectly. Moreover, optical properties of PTFE stay constant over a wide range of wavelengths, from UV down to near infrared. In this region, the relation of its regular transmittance to diffuse transmittance is negligibly small, so light transmitted through a diffuser (PTFE sheet) radiates like Lambert's cosine law. Thus PTFE enables cosinusoidal angular response for a detector measuring the power of optical radiation at a surface, e.g. in solar irradiance measurements.

- Certain types of bullets are coated with PTFE to reduce wear on firearms's rifling that harder projectiles would cause. PTFE itself does not give a projectile an armor-piercing property.

- Its high corrosion resistance makes PTFE useful in laboratory environments, where it is used for lining containers, as a coating for magnetic stirrers, and as tubing for highly corrosive chemicals such as hydrofluoric acid, which will dissolve glass containers. It is used in containers for storing fluoroantimonic acid, a superacid.

- PTFE tubes are used in gas-gas heat exchangers in gas cleaning of waste incinerators. Unit power capacity is typically several megawatts.

- PTFE is widely used as a thread seal tape in plumbing applications, largely replacing paste thread dope.

- PTFE membrane filters are among the most efficient industrial air filters. PTFE-coated filters are often used in dust collection systems to collect particulate matter from air streams in applications involving high temperatures and high particulate loads such as coal-fired power plants, cement production and steel foundries.

- PTFE grafts can be used to bypass stenotic arteries in peripheral vascular disease if a suitable autologous vein graft is not available.

- Many bicycle lubricants contain PTFE and are used on chains and other moving parts.

- PTFE can also be used for dental fillings, to isolate the contacts of the anterior tooth so the filling materials will not stick to the adjacent tooth.

- PTFE sheets are used in the production of butane hash oil due to its non-stick properties and resistance to non-polar solvents.

Safety

Pyrolysis of PTFE is detectable at 200 °C (392 °F), and it evolves several fluorocarbon gases and a sublimate. An animal study conducted in 1955 concluded that it is unlikely that these products would be generated in amounts significant to health at temperatures below 250 °C (482 °F).

While PTFE is stable and nontoxic at lower temperatures, it begins to deteriorate after the temperature of cookware reaches about 260 °C (500 °F), and decomposes above 350 °C (662 °F). The degradation by-products can be lethal to birds, and can cause flu-like symptoms in humans—see polymer fume fever.

Meat is usually fried between 204 and 232 °C (399 and 450 °F), and most oils start to smoke before a temperature of 260 °C (500 °F) is reached, but there are at least two cooking oils (refined safflower oil at 265 °C (510 °F) and avocado oil at 271 °C (520 °F)) that have a higher smoke point.

The Environmental Working Group recommends against using dental floss made with PTFE. They state that "Exposure to PFCs has been associated with kidney and testicular cancer, high cholesterol, abnormal thyroid hormone levels, pregnancy-induced hypertension and preeclampsia, obesity and low birth weight . . . . PFCs pollute water, are persistent in the environment and remain in the body for years. Leading manufacturers of PFCs have agreed to phase out some of these chemicals by the end of 2015, including PFOA, the most notorious, which used to be a key ingredient in making Teflon. Unfortunately, there's no evidence that the chemicals that have replaced PFOA are much safer."

PFOA

Perfluorooctanoic acid (PFOA, or C8) has been used as a surfactant in the emulsion polymerization of PTFE, although several manufacturers have entirely discontinued its use. PFOA persists indefinitely in the environment. It is a toxicant and carcinogen in animals. PFOA has been detected in the blood of more than 98% of the general US population in the low and sub-parts per billion range, and levels are higher in chemical plant employees and surrounding subpopulations. The general population has been exposed to PFOA through massive dumping of C8 waste into the ocean and near the Ohio River Valley. PFOA has been detected in industrial waste, stain resistant carpets, carpet cleaning liquids, house dust, microwave popcorn bags, water, food and Teflon cookware.

As a result of a class-action lawsuit and community settlement with DuPont, three epidemiologists conducted studies on the population surrounding a chemical plant that was exposed to PFOA at levels greater than in the general population. The studies concluded that there was probably an association between PFOA exposure and six health outcomes: kidney cancer, testicular cancer, ulcerative colitis, thyroid disease, hypercholesterolemia (high cholesterol), and pregnancy-induced hypertension.

Overall, PTFE cookware is considered an insignificant exposure pathway to PFOA.

Similar Polymers

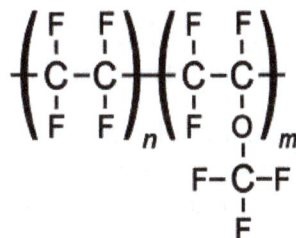

Teflon is also used as the trade name for a polymer with similar properties, perfluoroalkoxy polymer resin (PFA)

The Teflon trade name is also used for other polymers with similar compositions:

- Perfluoroalkoxy alkane (PFA)

- Fluorinated ethylene propylene (FEP)

These retain the useful PTFE properties of low friction and nonreactivity, but are more easily formable. For example, FEP is softer than PTFE and melts at 533 K (260 °C; 500 °F); it is also highly transparent and resistant to sunlight.

Polypropylene

Polypropylene (PP), also known as polypropene, is a thermoplastic polymer used in a wide variety of applications including packaging and labeling, textiles (e.g., ropes, thermal underwear and carpets), stationery, plastic parts and reusable containers of various types, laboratory equipment, loudspeakers, automotive components, and polymer banknotes. An addition polymer made from the monomer propylene, it is rugged and unusually resistant to many chemical solvents, bases and acids.

Polypropylene has a relatively slippery "low energy surface" that means that many common glues will not form adequate joints. Joining of polypropylene is often done using welding processes.

In 2013, the global market for polypropylene was about 55 million tonnes.

Chemical and Physical Properties

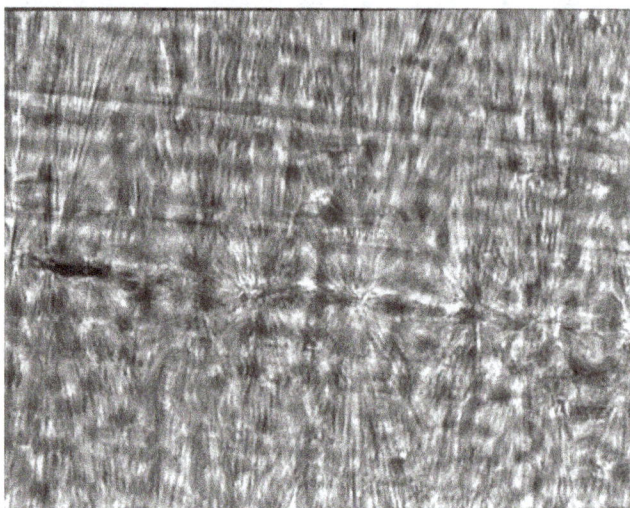

Micrograph of polypropylene

Polypropylene is in many aspects similar to polyethylene, especially in solution behaviour and electrical properties. The additionally present methyl group improves mechanical properties and thermal resistance, while the chemical resistance decreases. The properties of polypropylene depend on the molecular weight and molecular weight distribution, crystallinity, type and proportion of comonomer (if used) and the isotacticity. In isotactic polypropylene, for example, the CH3 groups are oriented on one side of the carbon backbone. This creates a greater degree of crystallinity and results in a stiffer material that is more resistant to creep than both atactic polypropylene and polyethylene.

Mechanical Properties

The density of PP is between 0.895 and 0.92 g/cm^3. Therefore, PP is the commodity plastic with the lowest density. With lower density, moldings parts with lower weight and more parts of a certain mass of plastic can be produced. Unlike polyethylene, crystalline and amorphous regions differ only slightly in their density. However, the density of polyethylene can significantly change with fillers.

The Young's modulus of PP is between 1300 and 1800 N/mm^2.

Polypropylene is normally tough and flexible, especially when copolymerized with ethylene. This allows polypropylene to be used as an engineering plastic, competing with materials such as acrylonitrile butadiene styrene (ABS). Polypropylene is reasonably economical.

Polypropylene has good resistance to fatigue.

Thermal Properties

The melting point of polypropylene occurs at a range, so a melting point is determined by finding the highest temperature of a differential scanning calorimetry chart. Perfectly isotactic PP has a melting point of 171 °C (340 °F). Commercial isotactic PP has a melting point that ranges from 160 to 166 °C (320 to 331 °F), depending on atactic material and crystallinity. Syndiotactic PP with a crystallinity of 30% has a melting point of 130 °C (266 °F). Below 0 °C, PP becomes brittle.

The thermal expansion of polypropylene is very large, but somewhat less than that of polyethylene.

Chemical Properties

Polypropylene is at room temperature resistant to fats and almost all organic solvents, apart from strong oxidants. Non-oxidizing acids and bases can be stored in containers made of PP. At elevated temperature, PP can be dissolved in nonpolarity solvents such as xylene, tetralin and decalin. Due to the tertiary carbon atom PP is chemically less resistant than PE.

Most commercial polypropylene is isotactic and has an intermediate level of crystallinity between that of low-density polyethylene (LDPE) and high-density polyethylene (HDPE). Isotactic & atactic polypropylene is soluble in P-xylene at 140 °C. Isotactic precipitates when the solution is cooled to 25 °C and atactic portion remains soluble in P-xylene.

The melt flow rate (MFR) or melt flow index (MFI) is a measure of molecular weight of polypropylene. The measure helps to determine how easily the molten raw material will flow during processing. Polypropylene with higher MFR will fill the plastic mold more easily during the injection or blow-molding production process. As the melt flow increases, however, some physical properties, like impact strength, will decrease.

There are three general types of polypropylene: homopolymer, random copolymer, and block copolymer. The comonomer is typically used with ethylene. Ethylene-propylene rubber or EPDM added to polypropylene homopolymer increases its low temperature impact strength. Randomly polymerized ethylene monomer added to polypropylene homopolymer decreases the polymer crystallinity, lowers the melting point and makes the polymer more transparent.

Degradation

Polypropylene is liable to chain degradation from exposure to heat and UV radiation such as that present in sunlight. Oxidation usually occurs at the tertiary carbon atom present in every repeat unit. A free radical is formed here, and then reacts further with oxygen, followed by chain scission to yield aldehydes and carboxylic acids. In external applications, it shows up as a network of fine cracks and crazes that become deeper and more severe with time of exposure. For external applications, UV-absorbing additives must be used. Carbon black also provides some protection from UV attack. The polymer can also be oxidized at high temperatures, a common problem during molding operations. Anti-oxidants are normally added to prevent polymer degradation. Microbial communities isolated from soil samples mixed with starch have been shown to be capable of degrading polypropylene. Polypropylene has been reported to degrade while in human body as implantable mesh devices. The degraded material forms a tree bark-like layer at the surface of mesh fibers.

Optical Properties

PP can be made translucent when uncolored but is not as readily made transparent as polystyrene, acrylic, or certain other plastics. It is often opaque or colored using pigments.

History

Phillips Petroleum chemists J. Paul Hogan and Robert L. Banks first polymerized propylene in 1951. Propylene was first polymerized to a crystalline isotactic polymer by Giulio Natta as well as by the German chemist Karl Rehn in March 1954. This pioneering discovery led to large-scale commercial production of isotactic polypropylene by the Italian firm Montecatini from 1957 onwards. Syndiotactic polypropylene was also first synthesized by Natta and his coworkers.

Polypropylene is the second most important plastic with revenues expected to exceed US$145 billion by 2019. The sales of this material are forecast to grow at a rate of 5.8% per year until 2021.

Synthesis

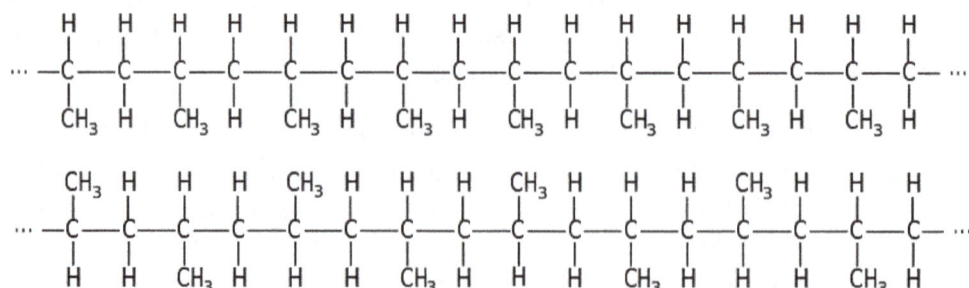

Short segments of polypropylene, showing examples of isotactic (above) and syndiotactic (below) tacticity.

An important concept in understanding the link between the structure of polypropylene and its properties is tacticity. The relative orientation of each methyl group (CH 3 in the figure) relative to the methyl groups in neighboring monomer units has a strong effect on the polymer's ability to form crystals.

A Ziegler-Natta catalyst is able to restrict linking of monomer molecules to a specific regular orientation, either isotactic, when all methyl groups are positioned at the same side with respect to the backbone of the polymer chain, or syndiotactic, when the positions of the methyl groups alternate. Commercially available isotactic polypropylene is made with two types of Ziegler-Natta catalysts. The first group of the catalysts encompasses solid (mostly supported) catalysts and certain types of soluble metallocene catalysts. Such isotactic macromolecules coil into a helical shape; these helices then line up next to one another to form the crystals that give commercial isotactic polypropylene many of its desirable properties.

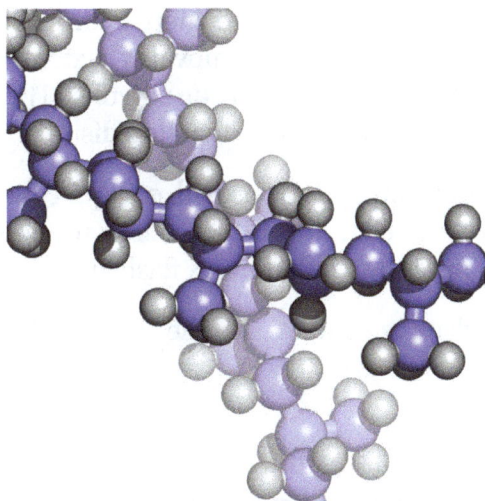

A ball-and-stick model of syndiotactic polypropylene.

Another type of metallocene catalysts produce syndiotactic polypropylene. These macromolecules also coil into helices (of a different type) and form crystalline materials.

When the methyl groups in a polypropylene chain exhibit no preferred orientation, the polymers are called atactic. Atactic polypropylene is an amorphous rubbery material. It can be produced commercially either with a special type of supported Ziegler-Natta catalyst or with some metallocene catalysts.

Modern supported Ziegler-Natta catalysts developed for the polymerization of propylene and other 1-alkenes to isotactic polymers usually use TiCl4 as an active ingredient and MgCl2 as a support. The catalysts also contain organic modifiers, either aromatic acid esters and diesters or ethers. These catalysts are activated with special cocatalysts containing an organoaluminum compound such as $Al(C_2H_5)_3$ and the second type of a modifier. The catalysts are differentiated depending on the procedure used for fashioning catalyst particles from $MgCl_2$ and depending on the type of organic modifiers employed during catalyst preparation and use in polymerization reactions. Two most important technological characteristics of all the supported catalysts are high productivity and a high fraction of the crystalline isotactic polymer they produce at 70–80 °C under standard polymerization conditions. Commercial synthesis of isotactic polypropylene is usually carried out either in the medium of liquid propylene or in gas-phase reactors.

Commercial synthesis of syndiotactic polypropylene is carried out with the use of a special class of metallocene catalysts. They employ bridged bis-metallocene complexes of the type bridge-$(Cp_1)(Cp_2)ZrCl_2$ where the first Cp ligand is the cyclopentadienyl group, the second Cp ligand is the fluorenyl group, and the bridge between the two Cp ligands is $-CH_2-CH_2-$, $>SiMe_2$, or $>SiPh_2$. These complexes are converted to polymerization catalysts by activating them with a special organoaluminum cocatalyst, methylaluminoxane (MAO).

Industrial Processes

Traditionally, three manufacturing processes are the most representative ways to produce polypropylene.

Hydrocarbon slurry or suspension: Uses a liquid inert hydrocarbon diluent in the reactor to facilitate transfer of propylene to the catalyst, the removal of heat from the system, the deactivation/removal of the catalyst as well as dissolving the atactic polymer. The range of grades that could be produced was very limited. (The technology has fallen into disuse).

Bulk (or bulk slurry): Uses liquid propylene instead of liquid inert hydrocarbon diluent. The polymer does not dissolve into a diluent, but rather rides on the liquid propylene. The formed polymer is withdrawn and any unreacted monomer is flashed off.

Gas phase: Uses gaseous propylene in contact with the solid catalyst, resulting in a fluidized-bed medium.

Manufacturing

Melting process of polypropylene can be achieved via extrusion and molding. Common extrusion methods include production of melt-blown and spun-bond fibers to form long rolls for future conversion into a wide range of useful products, such as face masks, filters, diapers and wipes.

The most common shaping technique is injection molding, which is used for parts such as cups, cutlery, vials, caps, containers, housewares, and automotive parts such as batteries. The related techniques of blow molding and injection-stretch blow molding are also used, which involve both extrusion and molding.

The large number of end-use applications for polypropylene are often possible because of the ability to tailor grades with specific molecular properties and additives during its manufacture. For example, antistatic additives can be added to help polypropylene surfaces resist dust and dirt. Many physical finishing techniques can also be used on polypropylene, such as machining. Surface treatments can be applied to polypropylene parts in order to promote adhesion of printing ink and paints.

Biaxially Oriented Polypropylene (BOPP)

When polypropylene film is extruded and stretched in both the machine direction and across machine direction it is called *biaxially oriented polypropylene*. Biaxial orientation increases strength and clarity. BOPP is widely used as a packaging material for packaging products such as snack foods, fresh produce and confectionery. It is easy to coat, print and laminate to give the required appearance and properties for use as a packaging material. This process is normally called converting. It is normally produced in large rolls which are slit on slitting machines into smaller rolls for use on packaging machines.

Development Trends

With the increase in the level of performance required for polypropylene quality in recent years, a variety of ideas and contrivances have been integrated into the production process for polypropylene.

There are roughly two directions for the specific methods. One is improvement of uniformity of the polymer particles produced using a circulation type reactor, and the other is improvement in the uniformity among polymer particles produced by using a reactor with a narrow retention time distribution.

Applications

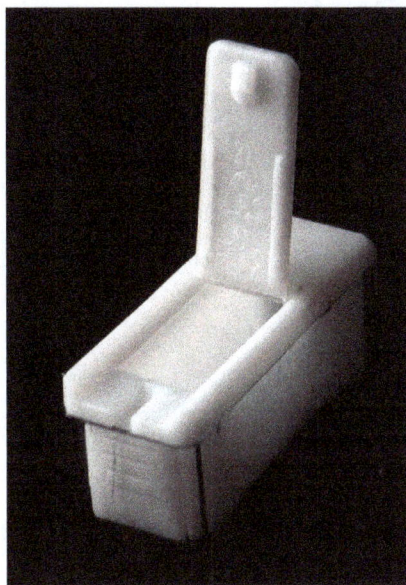

Polypropylene lid of a Tic Tacs box, with a living hinge and the resin identification code under its flap

As polypropylene is resistant to fatigue, most plastic living hinges, such as those on flip-top bottles, are made from this material. However, it is important to ensure that chain molecules are oriented across the hinge to maximise strength.

Very thin sheets (~2–20 μm) of polypropylene are used as a dielectric within certain high-performance pulse and low-loss RF capacitors.

Polypropylene is used in the manufacturing piping systems; both ones concerned with high-purity and ones designed for strength and rigidity (e.g. those intended for use in potable plumbing, hydronic heating and cooling, and reclaimed water). This material is often chosen for its resistance to corrosion and chemical leaching, its resilience against most forms of physical damage, including impact and freezing, its environmental benefits, and its ability to be joined by heat fusion rather than gluing.

Many plastic items for medical or laboratory use can be made from polypropylene because it can withstand the heat in an autoclave. Its heat resistance also enables it to be used as the manufacturing material of consumer-grade kettles. Food containers made from it will not melt in the dishwasher, and do not melt during industrial hot filling processes. For this reason, most plastic tubs for dairy products are polypropylene sealed with aluminum foil (both heat-resistant materials). After the product has cooled, the tubs are often given lids made of a less heat-resistant material, such as LDPE or polystyrene. Such containers provide a good hands-on example of the difference in modulus, since the rubbery (softer, more flexible) feeling of LDPE with respect to polypropylene of the same thickness is readily apparent. Rugged, translucent, reusable plastic containers made in a wide variety of shapes and sizes for consumers from various companies such as Rubbermaid and Sterilite are commonly made of polypropylene, although the lids are often made of somewhat more flexible LDPE so they can snap on to the container to close it. Polypropylene can also be made into disposable bottles to contain liquid, powdered, or similar consumer products, although HDPE and polyethylene terephthalate are commonly also used to make bottles. Plastic pails, car batteries, wastebaskets, pharmacy prescription bottles, cooler containers, dishes and pitchers are often made of polypropylene or HDPE, both of which commonly have rather similar appearance, feel, and properties at ambient temperature.

A polypropylene chair

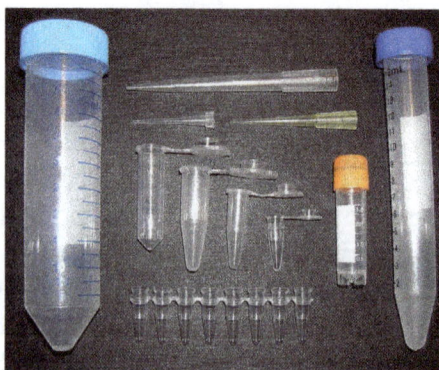

Polypropylene items for laboratory use, blue and orange closures are not made of polypropylene.

A common application for polypropylene is as biaxially oriented polypropylene (BOPP). These BOPP sheets are used to make a wide variety of materials including clear bags. When polypropylene is biaxially oriented, it becomes crystal clear and serves as an excellent packaging material for artistic and retail products.

Polypropylene, highly colorfast, is widely used in manufacturing carpets, rugs and mats to be used at home.

Polypropylene is widely used in ropes, distinctive because they are light enough to float in water. For equal mass and construction, polypropylene rope is similar in strength to polyester rope. Polypropylene costs less than most other synthetic fibers.

Polypropylene is also used as an alternative to polyvinyl chloride (PVC) as insulation for electrical cables for LSZH cable in low-ventilation environments, primarily tunnels. This is because it emits less smoke and no toxic halogens, which may lead to production of acid in high-temperature conditions.

Polypropylene is also used in particular roofing membranes as the waterproofing top layer of single-ply systems as opposed to modified-bit systems.

Polypropylene is most commonly used for plastic moldings, wherein it is injected into a mold while molten, forming complex shapes at relatively low cost and high volume; examples include bottle tops, bottles, and fittings.

It can also be produced in sheet form, widely used for the production of stationery folders, packaging, and storage boxes. The wide color range, durability, low cost, and resistance to dirt make it ideal as a protective cover for papers and other materials. It is used in Rubik's Cube stickers because of these characteristics.

The availability of sheet polypropylene has provided an opportunity for the use of the material by designers. The light-weight, durable, and colorful plastic makes an ideal medium for the creation of light shades, and a number of designs have been developed using interlocking sections to create elaborate designs.

Polypropylene sheets are a popular choice for trading card collectors; these come with pockets (nine for standard-size cards) for the cards to be inserted and are used to protect their condition and are meant to be stored in a binder.

Expanded polypropylene (EPP) is a foam form of polypropylene. EPP has very good impact characteristics due to its low stiffness; this allows EPP to resume its shape after impacts. EPP is extensively used in model aircraft and other radio controlled vehicles by hobbyists. This is mainly due to its ability to absorb impacts, making this an ideal material for RC aircraft for beginners and amateurs.

Polypropylene is used in the manufacture of loudspeaker drive units. Its use was pioneered by engineers at the BBC and the patent rights subsequently purchased by Mission Electronics for use in their Mission Freedom Loudspeaker and Mission 737 Renaissance loudspeaker.

Polypropylene fibres are used as a concrete additive to increase strength and reduce cracking and spalling. In the areas susceptible to earthquake, i.e., California, PP fibers are added with soils to improve the soils strength and damping when constructing the foundation of structures such as buildings, bridges, etc.

Polypropylene is used in polypropylene drums.

In June 2016, a study showed that a mixture of polypropylene and durable superoleophobic surfaces created by two engineers from Ohio State University can repel liquids such as shampoo and oil. This technology could make it easier to remove all the liquid contents from a polypropylene bottles, particularly those that have high surface tension such as shampoo or oil.

Clothing

Polypropylene is a major polymer used in nonwovens, with over 50% used for diapers or sanitary products where it is treated to absorb water (hydrophilic) rather than naturally repelling water (hydrophobic). Other interesting non-woven uses include filters for air, gas, and liquids in which the fibers can be formed into sheets or webs that can be pleated to form cartridges or layers that filter in various efficiencies in the 0.5 to 30 micrometre range. Such applications occur in houses as water filters or in air-conditioning-type filters. The high surface-area and naturally oleophilic polypropylene nonwovens are ideal absorbers of oil spills with the familiar floating barriers near oil spills on rivers.

Various polypropylene yarns and textiles

Polypropylene, or 'polypro', has been used for the fabrication of cold-weather base layers, such as long-sleeve shirts or long underwear. Polypropylene is also used in warm-weather clothing, in which it transports sweat away from the skin. More recently, polyester has replaced polypropylene in these applications in the U.S. military, such as in the ECWCS. Although polypropylene clothes are not easily flammable, they can melt, which may result in severe burns if the wearer is involved in an explosion or fire of any kind. Polypropylene undergarments are known for retaining body odors which are then difficult to remove. The current generation of polyester does not have this disadvantage.

Some fashion designers have adapted polypropylene to construct jewelry and other wearable items.

Medical

Its most common medical use is in the synthetic, nonabsorbable suture Prolene, manufactured by Ethicon Inc.

Polypropylene has been used in hernia and pelvic organ prolapse repair operations to protect the body from new hernias in the same location. A small patch of the material is placed over the spot of the hernia, below the skin, and is painless and rarely, if ever, rejected by the body. However, a polypropylene mesh will erode the tissue surrounding it over the uncertain period from days to years. Therefore, the FDA has issued several warnings on the use of polypropylene mesh medical kits for certain applications in pelvic organ prolapse, specifically when introduced in close proximity to the vaginal wall due to a continued increase in number of mesh-driven tissue erosions reported by patients over the past few years. Most recently, on 3 January 2012, the FDA ordered 35 manufacturers of these mesh products to study the side effects of these devices.

Initially considered inert, polypropylene has been found to degrade while in the body. The degraded material forms a bark-like shell on the mesh fibers and is prone to cracking.

EPP Model Aircraft

Since 2001, expanded polypropylene (EPP) foams have been gaining in popularity and in application as a structural material in hobbyist radio control model aircraft. Unlike expanded polystyrene foam (EPS) which is friable and breaks easily on impact, EPP foam is able to absorb kinetic impacts very well without breaking, retains its original shape, and exhibits memory form characteristics which allow it to return to its original shape in a short amount of time. In consequence, a radio-control model whose wings and fuselage are constructed from EPP foam is extremely resilient, and able to absorb impacts that would result in complete destruction of models made from lighter traditional materials, such as balsa or even EPS foams. EPP models, when covered with inexpensive fibreglass impregnated self-adhesive tapes, often exhibit much increased mechanical strength, in conjunction with a lightness and surface finish that rival those of models of the aforementioned types. EPP is also chemically highly inert, permitting the use of a wide variety of different adhesives. EPP can be heat molded, and surfaces can be easily finished with the use of cutting tools and abrasive papers. The principal areas of model making in which EPP has found great acceptance are the fields of:

- Wind-driven slope soarers

- Indoor electric powered profile electric models

- Hand launched gliders for small children

In the field of slope soaring, EPP has found greatest favour and use, as it permits the construction of radio-controlled model gliders of great strength and maneuverability. In consequence, the disciplines of slope combat (the active process of friendly competitors attempting to knock each other's planes out of the air by direct contact) and slope pylon racing have become commonplace, in direct consequence of the strength characteristics of the material EPP.

Building Construction

When the cathedral on Tenerife, La Laguna Cathedral, was repaired in 2002–2014, it turned out that the vaults and dome were in a rather bad condition. Therefore, these parts of the building were demolished, and replaced by constructions in polypropylene. This was reported as the first time this material was used in this scale in buildings.

Rope

Under the trade name Ulstron polypropylene rope is used to manufacture scoop nets for whitebait. It has also been used for sheets of yacht sails.

Recycling

Polypropylene is recyclable and has the number "5" as its resin identification code:

Repairing

Many objects are made with polypropylene precisely because it is resilient and resistant to most solvents and glues. Also, there are very few glues available specifically for gluing PP. However, solid PP objects not subject to undue flexing can be satisfactorily joined with a two part epoxy glue or using hot-glue guns. Preparation is important and it is often helpful to roughen the surface with a file, emery paper or other abrasive material to provide better anchorage for the glue. Also it is recommended to clean with mineral spirits or similar alcohol prior to gluing to remove any oils or other contamination. Some experimentation may be required. There are also some industrial glues available for PP, but these can be difficult to find, especially in a retail store.

PP can be melted using a speed welding technique. With speed welding, the plastic welder, similar to a soldering iron in appearance and wattage, is fitted with a feed tube for the plastic weld rod. The speed tip heats the rod and the substrate, while at the same time it presses the molten weld rod into position. A bead of softened plastic is laid into the joint, and the parts and weld rod fuse. With polypropylene, the melted welding rod must be "mixed" with the semi-melted base material being fabricated or repaired. A speed tip "gun" is essentially a soldering iron with a broad, flat tip that can be used to melt the weld joint and filler material to create a bond.

Health Concerns

The Environmental Working Group classifies PP as of low to moderate hazard. PP is dope-dyed, no water is used in its dyeing, in contrast with cotton.

In 2008, researchers in Canada asserted that quaternary ammonium biocides and oleamide were leaking out of certain polypropylene labware, affecting experimental results. As polypropylene is used in a wide number of food containers such as those for yogurt, Health Canada media spokesman Paul Duchesne said the department will be reviewing the findings to determine if steps are needed to protect consumers.

Polyurethane

Polyurethane (PUR and PU) is a polymer composed of organic units joined by carbamate (urethane) links. While most polyurethanes are thermosetting polymers that do not melt when heated, thermoplastic polyurethanes are also available.

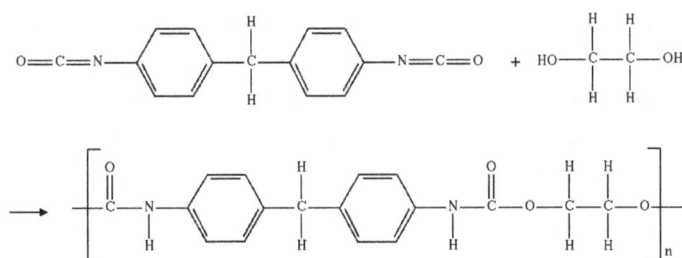

Polyurethane synthesis, wherein the urethane groups –NH–(C=O)–O– link the molecular units.

Polyurethane polymers are traditionally and most commonly formed by reacting a di- or polyisocyanate with a polyol. Both the isocyanates and polyols used to make polyurethanes contain, on average, two or more functional groups per molecule.

Some noteworthy recent efforts have been dedicated to minimizing the use of isocyanates to synthesize polyurethanes, because the isocyanates raise severe toxicity issues. Non-isocyanate based polyurethanes (NIPUs) have recently been developed as a new class of polyurethane polymers to mitigate health and environmental concerns.

Polyurethane products often are simply called "urethanes", but should not be confused with ethyl carbamate, which is also called urethane. Polyurethanes neither contain nor are produced from ethyl carbamate.

Polyurethanes are used in the manufacture of high-resilience foam seating, rigid foam insulation panels, microcellular foam seals and gaskets, durable elastomeric wheels and tires (such as roller coaster, escalator, shopping cart, elevator, and skateboard wheels), automotive suspension bushings, electrical potting compounds, high performance adhesives, surface coatings and surface sealants, synthetic fibers (e.g., Spandex), carpet underlay, hard-plastic parts (e.g., for electronic instruments), condoms, and hoses.

History

Otto Bayer and his coworkers at IG Farben in Leverkusen, Germany, first made polyurethanes in 1937. The new polymers had some advantages over existing plastics that were made by polymerizing olefins or by polycondensation, and were not covered by patents obtained by Wallace Carothers on polyesters. Early work focused on the production of fibres and flexible foams and PUs were applied on a limited scale as aircraft coating during World War II. Polyisocyanates became commercially available in 1952, and production of flexible polyurethane foam began in 1954 using toluene diisocyanate (TDI) and polyester polyols. These materials were also used to produce rigid foams, gum rubber, and elastomers. Linear fibers were produced from hexamethylene diisocyanate (HDI) and 1,4-butanediol (BDO).

In 1956 DuPont introduced polyether polyols, specifically poly(tetramethylene ether) glycol, and BASF and Dow Chemical started selling polyalkylene glycols in 1957. Polyether polyols were cheaper, easier to handle and more water-resistant than polyester polyols, and became more popular. Union Carbide and Mobay, a U.S. Monsanto/Bayer joint venture, also began making polyurethane chemicals. In 1960 more than 45,000 metric tons of flexible polyurethane foams were produced. The availability of chlorofluoroalkane blowing agents, inexpensive polyether polyols, and methylene diphenyl diisocyanate (MDI) allowed polyurethane rigid foams to be used as high-performance insulation materials. In 1967, urethane-modified polyisocyanurate rigid foams were introduced, offering even better thermal stability and flammability resistance. During the 1960s, automotive interior safety components, such as instrument and door panels, were produced by back-filling thermoplastic skins with semi-rigid foam.

In 1969, Bayer exhibited an all-plastic car in Düsseldorf, Germany. Parts of this car, such as the fascia and body panels, were manufactured using a new process called reaction injection molding (RIM), in which the reactants were mixed and then injected into a mold. The addition of fillers, such as milled glass, mica, and processed mineral fibres, gave rise to reinforced RIM (RRIM), which provided improvements in flexural modulus (stiffness), reduction in coefficient of thermal expansion and better thermal stability.This technology was used to make the first plastic-body automobile in the United States, the Pontiac Fiero, in 1983. Further increases in stiffness were obtained by incorporating pre-placed glass mats into the RIM mold cavity, also known broadly as resin injection molding, or structural RIM.

Starting in the early 1980s, water-blown microcellular flexible foams were used to mold gaskets for automotive panels and air filter seals, replacing PVC plastisol from automotive applications have greatly increased market share. Polyurethane foams are now used in high-temperature oil filter applications.

Polyurethane foam (including foam rubber) is sometimes made using small amounts of blowing agents to give less dense foam, better cushioning/energy absorption or thermal insulation. In the early 1990s, because of their impact on ozone depletion, the Montreal Protocol restricted the use of many chlorine-containing blowing agents, such as trichlorofluoromethane (CFC-11). By the late 1990s, blowing agents such as carbon dioxide, pentane, 1,1,1,2-tetrafluoroethane (HFC-134a) and 1,1,1,3,3-pentafluoropropane (HFC-245fa) were widely used in North America and the EU, although chlorinated blowing agents remained in use in many developing countries.

Chemistry

Polyurethanes are in the class of compounds called reaction polymers, which include epoxies, unsaturated polyesters, and phenolics. Polyurethanes are produced by reacting an isocyanate containing two or more isocyanate groups per molecule ($R-(N=C=O)_n$) with a polyol containing on average two or more hydroxyl groups per molecule ($R'-(OH)_n$) in the presence of a catalyst or by activation with ultraviolet light.

The properties of a polyurethane are greatly influenced by the types of isocyanates and polyols used to make it. Long, flexible segments, contributed by the polyol, give soft, elastic polymer. High amounts of crosslinking give tough or rigid polymers. Long chains and low crosslinking give a polymer that is very stretchy, short chains with lots of crosslinks produce a hard polymer while long chains and intermediate crosslinking give a polymer useful for making foam. The crosslinking present in polyurethanes means that the polymer consists of a three-dimensional network and molecular weight is very high. In some respects a piece of polyurethane can be regarded as one giant molecule. One consequence of this is that typical polyurethanes do not soften or melt when they are heated; they are thermosetting polymers. The choices available for the isocyanates and polyols, in addition to other additives and processing conditions allow polyurethanes to have the very wide range of properties that make them such widely used polymers.

Isocyanates are very reactive materials. This makes them useful in making polymers but also requires special care in handling and use. The aromatic isocyanates, diphenylmethane diisocyanate (MDI) or toluene diisocyanate (TDI) are more reactive than aliphatic isocyanates, such as hexamethylene diisocyanate (HDI) or isophorone diisocyanate (IPDI). Most of the isocyanates are difunctional, that is they have exactly two isocyanate groups per molecule. An important exception to this is polymeric diphenylmethane diisocyanate, which is a mixture of molecules with two, three, and four or more isocyanate groups. In cases like this the material has an average functionality greater than two, commonly 2.7.

Polyols are polymers in their own right and have on average two or more hydroxyl groups per molecule. Polyether polyols are mostly made by co-polymerizing ethylene oxide and propylene oxide with a suitable polyol precursor. Polyester polyols are made similarly to polyester polymers. The polyols used to make polyurethanes are not "pure" compounds since they are often mixtures of similar molecules with different molecular weights and mixtures of molecules that contain different numbers of hydroxyl groups, which is why the "average functionality" is often mentioned. Despite them being complex mixtures, industrial grade polyols have their composition sufficiently well controlled to produce polyurethanes having consistent properties. As mentioned earlier, it is the length of the polyol chain and the functionality that contribute much to the properties of the final polymer. Polyols used to make rigid polyurethanes have molecular weights in the hundreds, while those used to make flexible polyurethanes have molecular weights up to ten thousand or more.

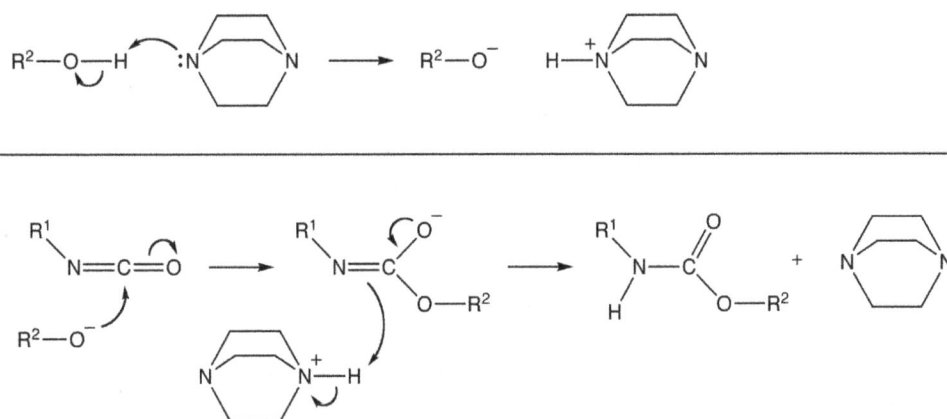

PU reaction mechanism catalyzed by a tertiary amine

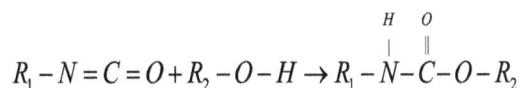

$$R_1-N=C=O+R_2-O-H \rightarrow R_1-\overset{H}{\underset{|}{N}}-\overset{O}{\overset{\|}{C}}-O-R_2$$

Generalized urethane reaction

The polymerization reaction makes a polymer containing the urethane linkage, −RNHCOOR′− and is catalyzed by tertiary amines, such as 1,4-diazabicyclo[2.2.2]octane (also called DABCO), and metallic compounds, such as dibutyltin dilaurate or bismuth octanoate. Alternatively, it can be promoted by ultraviolet light. This is often referred to as the gellation reaction or simply gelling.

If water is present in the reaction mixture (it is often added intentionally to make foams), the isocyanate reacts with water to form a urea linkage and carbon dioxide gas and the resulting polymer contains both urethane and urea linkages. This reaction is referred to as the blowing reaction and is catalyzed by tertiary amines like bis-(2-dimethylaminoethyl)ether.

A third reaction, particularly important in making insulating rigid foams is the isocyanate trimerization reaction, which is catalyzed by potassium octoate, for example.

One of the most desirable attributes of polyurethanes is their ability to be turned into foam. Making a foam requires the formation of a gas at the same time as the urethane polymerization (gellation) is occurring. The gas can be carbon dioxide, either generated by reacting isocyanate with water or added as a gas; it also be produced by boiling volatile liquids. In the latter case heat generated by the polymerization causes the liquids to vaporize. The liquids can be HFC-245fa (1,1,1,3,3-pentafluoropropane) and HFC-134a (1,1,1,2-tetrafluoroethane), and hydrocarbons such as n-pentane.

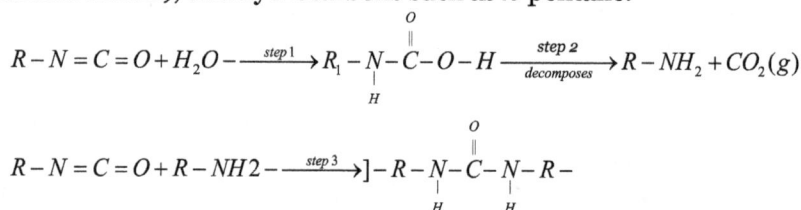

$$R-N=C=O+H_2O \xrightarrow{step 1} R_1-\underset{H}{N}-\overset{O}{\overset{\|}{C}}-O-H \xrightarrow[decomposes]{step 2} R-NH_2+CO_2(g)$$

$$R-N=C=O+R-NH2 \xrightarrow{step 3}]-R-\underset{H}{N}-\overset{O}{\overset{\|}{C}}-\underset{H}{N}-R-$$

Carbon dioxide gas formed by reacting water and isocyanate

The balance between gellation and blowing is sensitive to operating parameters including the concentrations of water and catalyst. The reaction to generate carbon dioxide involves water reacting with an isocyanate first forming an unstable carbamic acid, which then decomposes into carbon dioxide and an amine. The amine reacts with more isocyanate to give a substituted urea. Water has a very low molecular weight, so even though the weight percent of water may be small, the molar proportion of water may be high and considerable amounts of urea produced. The urea is not very soluble in the reaction mixture and tends to form separate "hard segment" phases consisting mostly of polyurea. The concentration and organization of these polyurea phases can have a significant impact on the properties of the polyurethane foam.

High-density microcellular foams can be formed without the addition of blowing agents by mechanically frothing or nucleating the polyol component prior to use.

Surfactants are used in polyurethane foams to emulsify the liquid components, regulate cell size, and stabilize the cell structure to prevent collapse and surface defects. Rigid foam surfactants are designed to produce very fine cells and a very high closed cell content. Flexible foam surfactants are designed to stabilize the reaction mass while at the same time maximizing open cell content to prevent the foam from shrinking.

An even more rigid foam can be made with the use of specialty trimerization catalysts which create cyclic structures within the foam matrix, giving a harder, more thermally stable structure, designated as polyisocyanurate foams. Such properties are desired in rigid foam products used in the construction sector.

Careful control of viscoelastic properties – by modifying the catalysts and polyols used – can lead to memory foam, which is much softer at skin temperature than at room temperature.

Foams can be either "closed-cell", where most of the original bubbles or cells remain intact, or "open-cell", where the bubbles have broken but the edges of the bubbles are stiff enough to retain their shape. Open-cell foams feel soft and allow air to flow through, so they are comfortable when used in seat cushions or mattresses. Closed-cell rigid foams are used as thermal insulation, for example in refrigerators.

Microcellular foams are tough elastomeric materials used in coverings of car steering wheels or shoe soles.

Raw Materials

The main ingredients to make a polyurethane are di- and triisocyanates and polyols. Other materials are added to aid processing the polymer or to modify the properties of the polymer.

Isocyanates

Isocyanates used to make polyurethane have two or more isocyanate groups on each molecule. The most commonly used isocyanates are the aromatic diisocyantes, toluene diisocyanate (TDI) and methylene diphenyl diisocyanate, MDI.

TDI and MDI are generally less expensive and more reactive than other isocyanates. Industrial grade TDI and MDI are mixtures of isomers and MDI often contains polymeric materials. They are used to make flexible foam (for example slabstock foam for mattresses or molded foams for car seats), rigid foam (for example insulating foam in refrigerators) elastomers (shoe soles, for example), and so on. The isocyanates may be modified by partially reacting them with polyols or introducing some other materials to reduce volatility (and hence toxicity) of the isocyanates, decrease their freezing points to make handling easier or to improve the properties of the final polymers.

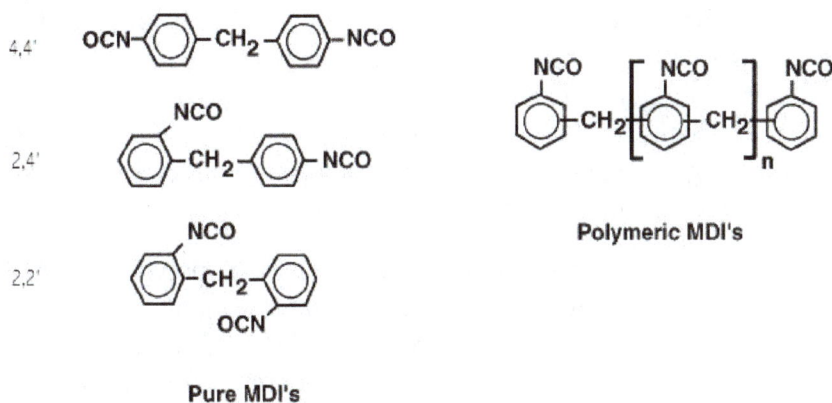

Pure MDI's

Polymeric MDI's

Aliphatic and cycloaliphatic isocyanates are used in smaller quantities, most often in coatings and other applications where color and transparency are important since polyurethanes made with aromatic isocyanates tend to darken on exposure to light. The most important aliphatic and cycloaliphatic isocyanates are 1,6-hexamethylene diisocyanate (HDI), 1-isocyanato-3-isocyanatomethyl-3,5,5-trimethyl-cyclohexane (isophorone diisocyanate, IPDI), and 4,4′-diisocyanato dicyclohexylmethane, (H_{12}MDI or hydrogenated MDI).

Polyols

Polyols can be polyether polyols, which are made by the reaction of epoxides with an active hydrogen containing compounds Polyester polyols are made by the polycondensation of multifunctional carboxylic acids and polyhydroxyl compounds. They can be further classified according to their end use. Higher molecular weight polyols (molecular weights from 2,000 to 10,000) are used to make more flexible polyurethanes while lower molecular weight polyols make more rigid products.

Polyols for flexible applications use low functionality initiators such as dipropylene glycol (f = 2), glycerine (f = 3), or a sorbitol/water solution (f = 2.75). Polyols for rigid applications use high functionality initiators such as sucrose (f = 8), sorbitol (f = 6), toluenediamine (f = 4), and Mannich bases (f = 4). Propylene oxide and/or ethylene oxide is added to the initiators until the desired molecular weight is achieved. The order of addition and the amounts of each oxide affect many polyol properties, such as compatibility, water-solubility, and reactivity. Polyols made with only propylene oxide are terminated with secondary hydroxyl groups and are less reactive than polyols capped with ethylene oxide, which contain primary hydroxyl groups. Graft polyols (also called filled polyols or polymer polyols) contain finely dispersed styrene–acrylonitrile, acrylonitrile, or polyurea (PHD) polymer solids chemically grafted to a high molecular weight polyether backbone. They are used to increase the load-bearing properties of low-density high-resiliency (HR) foam, as well as add toughness to microcellular foams and cast elastomers. Initiators such as ethylenediamine and triethanolamine are used to make low molecular weight rigid foam polyols that have built-in catalytic activity due to the presence of nitrogen atoms in the backbone. A special class of polyether polyols, poly(tetramethylene ether) glycols, which are made by polymerizing tetrahydrofuran, are used in high performance coating, wetting and elastomer applications.

Conventional polyester polyols are based on virgin raw materials and are manufactured by the direct polyesterification of high-purity diacids and glycols, such as adipic acid and 1,4-butanediol. Polyester polyols are usually more expensive and more viscous than polyether polyols, but they make polyurethanes with better solvent, abrasion, and cut resistance. Other polyester polyols are based on reclaimed raw materials. They are manufactured by transesterification (glycolysis) of recycled poly(ethyleneterephthalate) (PET) or dimethylterephthalate (DMT) distillation bottoms with glycols such as diethylene glycol. These low molecular weight, aromatic polyester polyols are used in rigid foam, and bring low cost and excellent flammability characteristics to polyisocyanurate (PIR) boardstock and polyurethane spray foam insulation.

Specialty polyols include polycarbonate polyols, polycaprolactone polyols, polybutadiene polyols, and polysulfide polyols. The materials are used in elastomer, sealant, and adhesive applications that require superior weatherability, and resistance to chemical and environmental attack. Natural oil polyols derived from castor oil and other vegetable oils are used to make elastomers, flexible bunstock, and flexible molded foam.

Co-polymerizing chlorotrifluoroethylene or tetrafluoroethylene with vinyl ethers containing hydroxyalkyl vinyl ether produces fluorinated (FEVE) polyols. Two-component fluorinated polyurethanes prepared by reacting FEVE fluorinated polyols with polyisocyanate have been used to make ambient cure paints and coatings. Since fluorinated polyurethanes contain a high percentage of fluorine–carbon bonds, which are the strongest bonds among all chemical bonds, fluorinated polyurethanes exhibit resistance to UV, acids, alkali, salts, chemicals, solvents, weathering, corrosion, fungi and microbial attack. These have been used for high performance coatings and paints.

Phosphorus-containing polyols are available that become chemically bonded to the polyurethane matrix for the use as flame retardants. This covalent linkage prevents migration and leaching of the organophosphorus compound.

Even polyols prepared from renewable sources like vegetable oils, derivatives of vegetable oil, sorbitol, cellulose, etc. are also reported for preparing polyurethane coatings.

Bio-derived Materials

Interest in sustainable "green" products raised some interest in polyols derived from vegetable oils. Many polyols are derived from renewable raw materials like vegetable oils. These oils include soybean, cotton seed and castor. Renewable source used to prepare polyols may be dimer fatty acid or fatty acid.

Some biobased and isocyanate-free polyurethanes exploit the reaction between polyamines and cyclic carbonates to produce polyhydroxurethanes.

Chain Extenders and Cross Linkers

Chain extenders (f = 2) and cross linkers ($f \geq 3$) are low molecular weight hydroxyl and amine terminated compounds that play an important role in the polymer morphology of polyurethane fibers, elastomers, adhesives, and certain integral skin and microcellular foams. The elastomeric properties of these materials are derived from the phase separation of the hard and soft copolymer segments of the polymer, such that the urethane hard segment domains serve as cross-links between the amorphous polyether (or polyester) soft segment domains. This phase separation occurs because the mainly nonpolar, low melting soft segments are incompatible with the polar, high melting hard segments. The soft segments, which are formed from high molecular weight polyols, are mobile and are normally present in coiled formation, while the hard segments, which are formed from the isocyanate and chain extenders, are stiff and immobile. Because the hard segments are covalently coupled to the soft segments, they inhibit plastic flow of the polymer chains, thus creating elastomeric resiliency. Upon mechanical deformation, a portion of the soft segments are stressed by uncoiling, and the hard segments become aligned in the stress direction. This reorientation of the hard segments and consequent powerful hydrogen bonding contributes to high tensile strength, elongation, and tear resistance values. The choice of chain extender also determines flexural, heat, and chemical resistance properties. The most important chain extenders are ethylene glycol, 1,4-butanediol (1,4-BDO or BDO), 1,6-hexanediol, cyclohexane dimethanol and hydroquinone bis(2-hydroxyethyl) ether (HQEE). All of these glycols form polyurethanes that phase separate well and form well defined hard segment domains, and are melt processable. They are all suitable for thermoplastic polyurethanes with the exception of ethylene glycol, since its derived bis-phenyl urethane undergoes unfavorable degradation at high hard segment levels. Diethanolamine and triethanolamine are used in flex molded foams to build firmness and add catalytic activity. Diethyltoluenediamine is used extensively in RIM, and in polyurethane and polyurea elastomer formulations.

Table of chain extenders and cross linkers				
Hydroxyl compounds – difunctional molecules				
	Mol. wt	Density (g/cm³)	m.p. (°C)	b.p. (°C)
Ethylene glycol	62.1	1.110	−13.4	197.4
Diethylene glycol	106.1	1.111	−8.7	245.5
Triethylene glycol	150.2	1.120	−7.2	287.8
Tetraethylene glycol	194.2	1.123	−9.4	325.6
Propylene glycol	76.1	1.032	Supercools	187.4
Dipropylene glycol	134.2	1.022	Supercools	232.2
Tripropylene glycol	192.3	1.110	Supercools	265.1
1,3-Propanediol	76.1	1.060	−28	210
1,3-Butanediol	92.1	1.005	−	207.5
1,4-Butanediol	92.1	1.017	20.1	235
Neopentyl glycol	104.2	−	130	206
1,6-Hexanediol	118.2	1.017	43	250
1,4-Cyclohexanedimethanol	−	−	−	−
HQEE	−	−	−	−
Ethanolamine	61.1	1.018	10.3	170
Diethanolamine	105.1	1.097	28	271
Methyldiethanolamine	119.1	1.043	−21	242
Phenyldiethanolamine	181.2	−	58	228
Hydroxyl compounds – trifunctional molecules				
	Mol. wt	Density (g/cm³)	m.p. (°C)	b.p. (°C)
Glycerol	92.1	1.261	18.0	290
Trimethylolpropane	−	−	−	−
1,2,6-Hexanetriol	−	−	−	−
Triethanolamine	149.2	1.124	21	−
Hydroxyl compounds – tetrafunctional molecules				
	Mol. wt	Density (g/cm³)	m.p. (°C)	b.p. (°C)
Pentaerythritol	136.2	−	260.5	−
N,N,N′,N′-Tetrakis (2-hydroxypropyl) ethylenediamine	−	−	−	−
Amine compounds – difunctional molecules				
	Mol. wt	Density (g/cm³)	m.p. (°C)	b.p. (°C)
Diethyltoluenediamine	178.3	1.022	−	308
Dimethylthiotoluenediamine	214.0	1.208	−	−

Catalysts

Polyurethane catalysts can be classified into two broad categories, basic and acidic amine. Tertiary amine catalysts function by enhancing the nucleophilicity of the diol component. Alkyl tin carboxylates, oxides and mercaptides oxides function as mild Lewis acids in accelerating the formation of polyurethane. As bases, traditional amine catalysts include triethylenediamine (TEDA, also called DABCO, 1,4-diazabicyclo[2.2.2]octane), dimethylcyclohexylamine (DMCHA), and dimethylethanolamine (DMEA). A typical Lewis acidic catalyst is dibutyltin dilaurate. The process is highly sensitive to the nature of the catalyst and is also known to be autocatalytic.

Factors affecting catalyst selection include balancing three reactions: urethane (polyol+isocyanate, or gel) formation, the urea (water+isocyanate, or "blow") formation, or the isocyanate trimerization reaction (e.g., using potassium acetate, to form isocyanurate rings). A variety of specialized catalysts have been developed.

Surfactants

Surfactants are used to modify the characteristics of both foam and non-foam polyurethane polymers. They take the form of polydimethylsiloxane-polyoxyalkylene block copolymers, silicone oils, nonylphenol ethoxylates, and other organic compounds. In foams, they are used to emulsify the liquid components, regulate cell size, and stabilize the cell structure to prevent collapse and sub-surface voids. In non-foam applications they are used as air release and antifoaming agents, as wetting agents, and are used to eliminate surface defects such as pin holes, orange peel, and sink marks.

Production

Polyurethanes are produced by mixing two or more liquid streams. The polyol stream contains catalysts, surfactants, blowing agents and so on. The two components are referred to as a polyurethane system, or simply a system. The isocyanate is commonly referred to in North America as the 'B-side' or just the 'iso'. The blend of polyols and other additives is commonly referred to as the 'A-side' or as the 'poly'. This mixture might also be called a 'resin' or 'resin blend'. In Europe the meanings for 'A-side' and 'B-side' are reversed. Resin blend additives may include chain extenders, cross linkers, surfactants, flame retardants, blowing agents, pigments, and fillers. Polyurethane can be made in a variety of densities and hardnesses by varying the isocyanate, polyol or additives.

Health and safety

Fully reacted polyurethane polymer is chemically inert. No exposure limits have been established in the U.S. by OSHA (Occupational Safety and Health Administration) or ACGIH (American Conference of Governmental Industrial Hygienists). It is not regulated by OSHA for carcinogenicity.

Polyurethane polymer is a combustible solid and can be ignited if exposed to an open flame. Decomposition from fire can produce mainly carbon monoxide, and trace nitrogen oxides and hydrogen cyanide. Because of the flammability of the material, it has to be treated with flame retardants (at least in case of furniture), almost all of which are considered harmful.

Liquid resin blends and isocyanates may contain hazardous or regulated components. Isocyanates are known skin and respiratory sensitizers. Additionally, amines, glycols, and phosphate present in spray polyurethane foams present risks.

Exposure to chemicals that may be emitted during or after application of polyurethane spray foam (such as isocyanates) are harmful to human health and therefore special precautions are required during and after this process.

In the United States, additional health and safety information can be found through organizations such as the Polyurethane Manufacturers Association (PMA) and the Center for the Polyurethanes Industry (CPI), as well as from polyurethane system and raw material manufacturers. Regulatory information can be found in the Code of Federal Regulations Title 21 (Food and Drugs) and Title 40 (Protection of the Environment). In Europe, health and safety information is available from ISOPA, the European Diisocyanate and Polyol Producers Association.

Manufacturing

The methods of manufacturing polyurethane finished goods range from small, hand pour piece-part operations to large, high-volume bunstock and boardstock production lines. Regardless of the end-product, the manufacturing principle is the same: to meter the liquid isocyanate and resin blend at a specified stoichiometric ratio, mix them together until a homogeneous blend is obtained, dispense the reacting liquid into a mold or on to a surface, wait until it cures, then demold the finished part.

Dispensing Equipment

A high pressure polyurethane dispense unit, showing control panel, high pressure pump, integral day tanks, and hydraulic drive unit.	A high pressure mix head, showing simple controls. Front view.	A high pressure mix head, showing material supply and hydraulic actuator lines. Rear view.

Although the capital outlay can be high, it is desirable to use a meter-mix or dispense unit for even low-volume production operations that require a steady output of finished parts. Dispense equipment consists of material holding (day) tanks, metering pumps, a mix head, and a control unit. Often, a conditioning or heater-chiller unit is added to control material temperature in order to improve mix efficiency, cure rate, and to reduce process variability. Choice of dispense equipment components depends on shot size, throughput, material characteristics such as viscosity and filler content, and process control. Material day tanks may be single to hundreds of gallons in size, and may be supplied directly from drums, IBCs (intermediate bulk containers, such as totes), or bulk storage tanks. They may incorporate level sensors, conditioning jackets, and mixers. Pumps can be sized to meter in single grams per second up to hundreds of pounds per minute. They can be rotary, gear, or piston pumps, or can be specially hardened lance pumps to meter liquids containing highly abrasive fillers such as chopped or hammer milled glass fibres and wollastonite.

A low pressure mix head with calibration chamber installed, showing material supply and air actuator lines.	Low pressure mix head components, including mix chambers, conical mixers, and mounting plates.	5-gallon (20-liter) material day tanks for supplying a low pressure dispense unit.

The pumps can drive low-pressure (10 to 30 bar, 1 to 3 MPa) or high-pressure (125 to 250 bar, 12.5 to 25.0 MPa) dispense systems. Mix heads can be simple static mix tubes, rotary element mixers, low-pressure dynamic mixers, or high-pressure hydraulically actuated direct impingement mixers. Control units may have basic on/off and dispense/stop switches, and analogue pressure and temperature gauges, or may be computer controlled with flow meters to electronically calibrate mix ratio, digital temperature and level sensors, and a full suite of statistical process control software. Add-ons to dispense equipment include nucleation or gas injection units, and third or fourth stream capability for adding pigments or metering in supplemental additive packages.

Tooling

Distinct from pour-in-place, bun and boardstock, and coating applications, the production of piece parts requires tooling to contain and form the reacting liquid. The choice of mold-making material is dependent on the expected number of uses to end-of-life (EOL), molding pressure, flexibility, and heat transfer characteristics.

RTV silicone is used for tooling that has an EOL in the thousands of parts. It is typically used for molding rigid foam parts, where the ability to stretch and peel the mold around undercuts is needed. The heat transfer characteristic of RTV silicone tooling is poor. High-performance, flexible polyurethane elastomers are also used in this way.

Epoxy, metal-filled epoxy, and metal-coated epoxy is used for tooling that has an EOL in the tens of thousands of parts. It is typically used for molding flexible foam cushions and seating, integral skin and microcellular foam padding, and shallow-draft RIM bezels and fascia. The heat transfer characteristic of epoxy tooling is fair; the heat transfer characteristic of metal-filled and metal-coated epoxy is good. Copper tubing can be incorporated into the body of the tool, allowing hot water to circulate and heat the mold surface.

Aluminum is used for tooling that has an EOL in the hundreds of thousands of parts. It is typically used for molding microcellular foam gasketing and cast elastomer parts, and is milled or extruded into shape.

Mirror-finish stainless steel is used for tooling that imparts a glossy appearance to the finished part. The heat transfer characteristic of metal tooling is excellent.

Finally, molded or milled polypropylene is used to create low-volume tooling for molded gasket applications. Instead of many expensive metal molds, low-cost plastic tooling can be formed from a single metal master, which also allows greater design flexibility. The heat transfer characteristic of polypropylene tooling is poor, which must be taken into consideration during the formulation process.

Applications

In 2007, the global consumption of polyurethane raw materials was above 12 million metric tons, the average annual growth rate is about 5%. Revenues generated with PUR on the global market are expected to rise to approximately US$80 billion by 2020.

Effects of Visible Light

Polyurethanes, especially those made using aromatic isocyanates, contain chromophores that interact with light. This is of particular interest in the area of polyurethane coatings, where light stability is a critical factor and is the main reason that aliphatic isocyanates are used in making polyurethane coatings. When PU foam, which is made using aromatic isocyanates, is exposed to visible light, it discolors, turning from off-white to yellow to reddish brown. It has been generally accepted that apart from yellowing, visible light has little effect on foam properties. This is especially the case if the yellowing happens on the outer portions of a large foam, as the deterioration of properties in the outer portion has little effect on the overall bulk properties of the foam itself.

It has been reported that exposure to visible light can affect the variability of some physical property test results.

Higher-energy UV radiation promotes chemical reactions in foam, some of which are detrimental to the foam structure.

Biodegradation

Two species of the Ecuadorian fungus *Pestalotiopsis* are capable of biodegrading polyurethane in aerobic and anaerobic conditions such as found at the bottom of landfills. Degradation of polyurethane items at museums has been reported. Polyester-type polyurethanes are more easily biodegraded by fungus than polyether-type.

References

- Koltzenburg, Sebastian; Maskos, Michael; Nuyken, Oskar (2014). Polymere: Synthese, Eigenschaften und Anwendungen (1 ed.). Springer Spektrum. p. 406. ISBN 978-3-642-34773-3.

- Elsner, Peter; Eyerer, Peter; Hirth, Thomas (2012). Domininghaus - Kunststoffe (8. ed.). Berlin Heidelberg: Springer-Verlag. p. 224. ISBN 978-3-642-16173-5.

- Hirth, von Hans Domininghaus; herausgegeben von Peter Elsner, Peter Eyerer, Thomas (2012). Kunststoffe Eigenschaften und Anwendungen (8. ed.). Berlin, Heidelberg: Springer Berlin Heidelberg. ISBN 978-3-642-16173-5.

- Baur, Erwin; Osswald, Tim A. (October 2013). Saechtling Kunststoff Taschenbuch. p. 443. ISBN 978-3-446-43729-6. Vorschau auf kunststoffe.de

- Pascu, Cornelia Vasile: Mihaela (2005). Practical guide to polyethylene ([Online-Ausg.]. ed.). Shawbury: Rapra Technology Ltd. ISBN 1859574939.

- Kaiser, Wolfgang (2011). Kunststoffchemie für Ingenieure von der Synthese bis zur Anwendung (3. ed.). München: Hanser. ISBN 978-3-446-43047-1.

- "A Guide to IUPAC Nomenclature of Organic Compounds (Recommendations 1993) IUPAC, Commission on Nomenclature of Organic Chemistry". Blackwell Scientific Publications. 1993. ISBN 0632037024. Retrieved 20 February 2014.

- Rhodes, Richard (1986). The Making of the Atomic Bomb. New York, New York: Simon and Schuster. p. 494. ISBN 0-671-65719-4. Retrieved 31 October 2010.

- John W. Nicholson (2011). The Chemistry of Polymers (4, Revised ed.). Royal Society of Chemistry. p. 50. ISBN 9781849733915. Retrieved 10 September 2013.

- Maier, Clive; Calafut, Teresa (1998). Polypropylene: the definitive user's guide and databook. William Andrew. p. 14. ISBN 978-1-884207-58-7.

- Kaiser, Wolfgang (2011). Kunststoffchemie für Ingenieure von der Synthese bis zur Anwendung (3. ed.). München: Hanser. ISBN 978-3-446-43047-1.

Synthesis of Polymers

Polymerization has a number of forms. Some of these are step-growth polymerization, chain-growth polymerization and reversible-deactivation polymerization. The forms of chain growth polymerization such as, anionic addition polymerization, cationic polymerization and living polymerization are also explained in the following chapter.

Polymerization

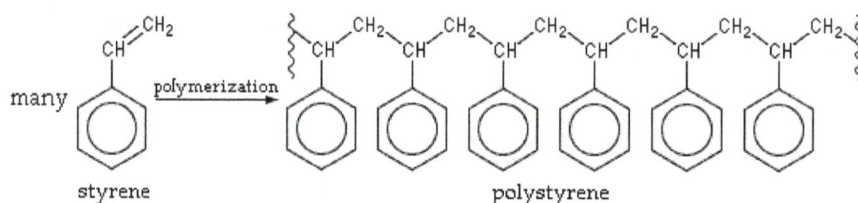

An example of alkene polymerization, in which each styrene monomer's double bond reforms as a single bond plus a bond to another styrene monomer. The product is polystyrene.

In polymer chemistry, polymerization is a process of reacting monomer molecules together in a chemical reaction to form polymer chains or three-dimensional networks. There are many forms of polymerization and different systems exist to categorize them.

Introduction

Homopolymers

$$A + A + A + A... \rightarrow AAAA...$$

Copolymers

$$A + B + A + B... \rightarrow ABAB...$$

In chemical compounds, polymerization occurs via a variety of reaction mechanisms that vary in complexity due to functional groups present in reacting compounds and their inherent steric effects. In more straightforward polymerization, alkenes, which are relatively stable due to σ bonding between carbon atoms, form polymers through relatively simple radical reactions; in contrast, more complex reactions such as those that involve substitution at the carbonyl group require more complex synthesis due to the way in which reacting molecules polymerize. Alkanes can also be polymerized, but only with the help of strong acids.

As alkenes can be formed in somewhat straightforward reaction mechanisms, they form useful compounds such as polyethylene and polyvinyl chloride (PVC) when undergoing radical reactions, which are produced in high tonnages each year due to their usefulness in manufacturing processes of commercial products, such as piping, insulation and packaging. In general, polymers such as PVC are referred to as "homopolymers," as they consist of repeated long chains or structures of the same monomer unit, whereas polymers that consist of more than one molecule are referred to as copolymers (or co-polymers).

Other monomer units, such as formaldehyde hydrates or simple aldehydes, are able to polymerize themselves at quite low temperatures (ca. −80 °C) to form trimers; molecules consisting of 3 monomer units, which can cyclize to form ring cyclic structures, or undergo further reactions to form tetramers, or 4 monomer-unit compounds. Further compounds either being referred to as oligomers in smaller molecules. Generally, because formaldehyde is an exceptionally reactive electrophile it allows nucleophillic addition of hemiacetal intermediates, which are in general short-lived and relatively unstable "mid-stage" compounds that react with other molecules present to form more stable polymeric compounds.

Polymerization that is not sufficiently moderated and proceeds at a fast rate can be very hazardous. This phenomenon is known as hazardous polymerization and can cause fires and explosions.

Step-growth

Step-growth polymers are defined as polymers formed by the stepwise reaction between functional groups of monomers, usually containing heteroatoms such as nitrogen or oxygen. Most step-growth polymers are also classified as condensation polymers, but not all step-growth polymers (like polyurethanes formed from isocyanate and alcohol bifunctional monomers) release condensates; in this case, we talk about addition polymers. Step-growth polymers increase in molecular weight at a very slow rate at lower conversions and reach moderately high molecular weights only at very high conversion (i.e., >95%).

To alleviate inconsistencies in these naming methods, adjusted definition for condensation and addition polymers have been developed. A condensation polymer is defined as a polymer that involves loss of small molecules during its synthesis, or contains heteroatoms as part of its backbone chain, or its repeat unit does not contain all the atoms present in the hypothetical monomer to which it can be degraded.

Chain-growth

Chain-growth polymerization (or addition polymerization) involves the linking together of molecules incorporating double or triple carbon-carbon bonds. These unsaturated *monomers* (the identical molecules that make up the polymers) have extra internal bonds that are able to break and link up with other monomers to form a repeating chain, whose backbone typically contains only carbon atoms. Chain-growth polymerization is involved in the manufacture of polymers such as polyethylene, polypropylene, and polyvinyl chloride (PVC). A special case of chain-growth polymerization leads to living polymerization.

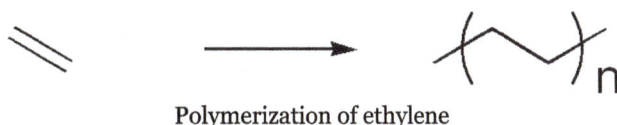

Polymerization of ethylene

In the radical polymerization of ethylene, its π bond is broken, and the two electrons rearrange to create a new propagating center like the one that attacked it. The form this propagating center takes depends on the specific type of addition mechanism. There are several mechanisms through which this can be initiated. The free radical mechanism is one of the first methods to be used. Free radicals are very reactive atoms or molecules that have unpaired electrons. Taking the polymerization of ethylene as an example, the free radical mechanism can be divided into three stages: chain initiation, chain propagation, and chain termination.

Free radical addition polymerization of ethylene must take place at high temperatures and pressures, approximately 300 °C and 2000 atm. While most other free radical polymerizations do not require such extreme temperatures and pressures, they do tend to lack control. One effect of this lack of control is a high degree of branching. Also, as termination occurs randomly, when two chains collide, it is impossible to control the length of individual chains. A newer method of polymerization similar to free radical, but allowing more control involves the Ziegler-Natta catalyst, especially with respect to polymer branching.

Other forms of chain growth polymerization include cationic addition polymerization and anionic addition polymerization. While not used to a large extent in industry yet due to stringent reaction conditions such as lack of water and oxygen, these methods provide ways to polymerize some monomers that cannot be polymerized by free radical methods such as polypropylene. Cationic and anionic mechanisms are also more ideally suited for living polymerizations, although free radical living polymerizations have also been developed.

Esters of acrylic acid contain a carbon-carbon double bond which is conjugated to an ester group. This allows the possibility of both types of polymerization mechanism. An acrylic ester by itself can undergo chain-growth polymerization to form a homopolymer with a carbon-carbon backbone, such as poly(methyl methacrylate). Also, however, certain acrylic esters can react with diamine monomers by nucleophilic conjugate addition of amine groups to acrylic C=C bonds. In this case the polymerization proceeds by step-growth and the products are poly(beta-amino ester) copolymers, with backbones containing nitrogen (as amine) and oxygen (as ester) as well as carbon.

Physical Polymer Reaction Engineering

To produce a high-molecular-weight, uniform product, various methods are employed to better control the initiation, propagation, and termination rates during chain polymerization and also to remove excess concentrated heat during these exothermic reactions compared to polymerization of the pure monomer (also referred to as bulk polymerization). These include emulsion polymerization, solution polymerization, suspension polymerization, and precipitation polymerization. Although the polymer polydispersity and molecular weight may be improved, these methods may introduce additional processing requirements to isolate the product from a solvent.

Photopolymerization

Most photopolymerization reactions are chain-growth polymerizations which are initiated by the absorption of visible or ultraviolet light. The light may be absorbed either directly by the reactant monomer (*direct* photopolymerization), or else by a *photosensitizer* which absorbs the light and then transfers energy to the monomer. In general only the initiation step differs from that of the ordinary thermal polymerization of the same monomer; subsequent propagation, termination and chain transfer steps are unchanged. In step-growth photopolymerization, absorption of light triggers an addition (or condensation) reaction between two comonomers that do not react without light. A propagation cycle is not initiated because each growth step requires the assistance of light.

Photopolymerization can be used as a photographic or printing process, because polymerization only occurs in regions which have been exposed to light. Unreacted monomer can be removed from unexposed regions, leaving a relief polymeric image. Several forms of 3D printing—including layer-by-layer stereolithography and two-photon absorption 3D photopolymerization—use photopolymerization.

Macromolecule

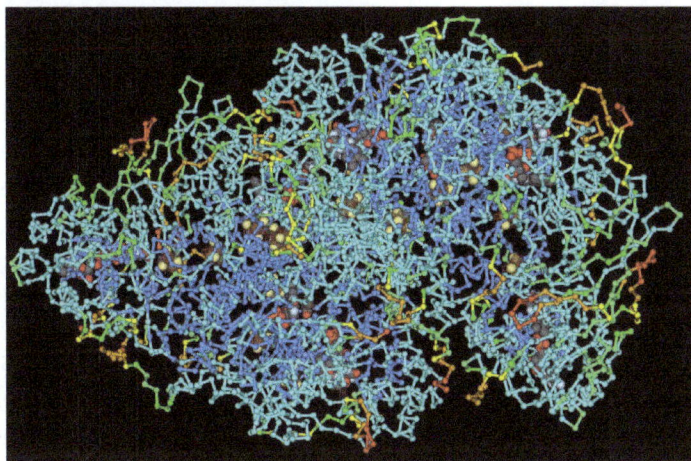

Chemical structure of a polypeptide macromolecule

A macromolecule is a very large molecule, such as protein, commonly created by polymerization of smaller subunits (monomers). They are typically composed of thousands of atoms or more. The most common macromolecues in biochemistry are biopolymers (nucleic acids, proteins, carbohydrates and polyphenols) and large non-polymeric molecules (such as lipids and macrocycles). Synthetic macromolecules include common plastics and synthetic fibres as well as experimental materials such as carbon nanotubes.

Definition

The term *macromolecule* (*macro-* + *molecule*) was coined by Nobel laureate Hermann Staudinger in the 1920s, although his first relevant publication on this field only mentions *high molecular compounds* (in excess of 1,000 atoms). At that time the phrase *polymer*, as introduced by Berzelius in 1833, had a different meaning from that of today: it simply was another form of isomerism for example with benzene and acetylene and had little to do with size.

Usage of the term to describe large molecules varies among the disciplines. For example, while biology refers to macromolecules as the four large molecules comprising living things, in chemistry, the term may refer to aggregates of two or more molecules held together by intermolecular forces rather than covalent bonds but which do not readily dissociate.

According to the standard IUPAC definition, the term *macromolecule* as used in polymer science refers only to a single molecule. For example, a single polymeric molecule is appropriately described as a "macromolecule" or "polymer molecule" rather than a "polymer", which suggests a substance composed of macromolecules.

Because of their size, macromolecules are not conveniently described in terms of stoichiometry alone. The structure of simple macromolecules, such as homopolymers, may be described in terms of the individual monomer subunit and total molecular mass. Complicated biomacromolecules, on the other hand, require multi-faceted structural description such as the hierarchy of structures used to describe proteins. In British English, the word "macromolecule" tends to be called "high polymer".

Properties

Macromolecules often have unusual physical properties that do not occur for smaller molecules.

For example, DNA in a solution can be broken simply by sucking the solution through an ordinary straw because the physical forces on the molecule can overcome the strength of its covalent bonds. The 1964 edition of Linus Pauling's *College Chemistry* asserted that DNA in nature is never longer than about 5,000 base pairs. This error arose because biochemists were inadvertently breaking their samples into fragments. In fact, the DNA of chromosomes can be hundreds of millions of base pairs long, packaged into chromatin.

Another common macromolecular property that does not characterize smaller molecules is their relative insolubility in water and similar solvents, instead forming colloids. Many require salts or particular ions to dissolve in water. Similarly, many proteins will denature if the solute concentration of their solution is too high or too low.

High concentrations of macromolecules in a solution can alter the rates and equilibrium constants of the reactions of other macromolecules, through an effect known as macromolecular crowding. This comes from macromolecules excluding other molecules from a large part of the volume of the solution, thereby increasing the effective concentrations of these molecules.

Linear Biopolymers

All living organisms are dependent on three essential biopolymers for their biological functions: DNA, RNA and Proteins. Each of these molecules is required for life since each plays a distinct, indispensable role in the cell. The simple summary is that DNA makes RNA, and then RNA makes proteins.

DNA, RNA and proteins all consist of a repeating structure of related building blocks (nucleotides in the case of DNA and RNA, amino acids in the case of proteins). In general, they are all unbranched polymers, and so can be represented in the form of a string. Indeed, they can be viewed as a string of beads, with each bead representing a single nucleotide or amino acid monomer linked together through covalent chemical bonds into a very long chain.

In most cases, the monomers within the chain have a strong propensity to interact with other amino acids or nucleotides. In DNA and RNA, this can take the form of Watson-Crick base pairs (G-C and A-T or A-U), although many more complicated interactions can and do occur.

Structural Features

	DNA	RNA	Proteins
Encodes genetic information	Yes	Yes	No
Catalyzes biological reactions	No	Yes	Yes
Building blocks (type)	Nucleotides	Nucleotides	Amino acids
Building blocks (number)	4	4	20
Strandedness	Double	Single	Single
Structure	Double helix	Complex	Complex
Stability to degradation	High	Variable	Variable
Repair systems	Yes	No	No

Because of the double-stranded nature of DNA, essentially all of the nucleotides take the form of Watson-Crick base pairs between nucleotides on the two complementary strands of the double-helix.

In contrast, both RNA and proteins are normally single-stranded. Therefore, they are not constrained by the regular geometry of the DNA double helix, and so fold into complex three-dimensional shapes dependent on their sequence. These different shapes are responsible for many of the common properties of RNA and proteins, including the formation of specific binding pockets, and the ability to catalyse biochemical reactions.

DNA is Optimised for Encoding Information

DNA is an information storage macromolecule that encodes the complete set of instructions (the genome) that are required to assemble, maintain, and reproduce every living organism.

DNA and RNA are both capable of encoding genetic information, because there are biochemical mechanisms which read the information coded within a DNA or RNA sequence and use it to generate a specified protein. On the other hand, the sequence information of a protein molecule is not used by cells to functionally encode genetic information.

DNA has three primary attributes that allow it to be far better than RNA at encoding genetic information. First, it is normally double-stranded, so that there are a minimum of two copies of the information encoding each gene in every cell. Second, DNA has a much greater stability against breakdown than does RNA, an attribute primarily associated with the absence of the 2'-hydroxyl group within every nucleotide of DNA. Third, highly sophisticated DNA surveillance and repair systems are present which monitor damage to the DNA and repair the sequence when necessary. Analogous systems have not evolved for repairing damaged RNA molecules. Consequently, chromosomes can contain many billions of atoms, arranged in a specific chemical structure.

Proteins are Optimised for Catalysis

Proteins are functional macromolecules responsible for catalysing the biochemical reactions that sustain life. Proteins carry out all functions of an organism, for example photosynthesis, neural function, vision, and movement.

The single-stranded nature of protein molecules, together with their composition of 20 or more different amino acid building blocks, allows them to fold in to a vast number of different three-dimensional shapes, while providing binding pockets through which they can specifically interact with all manner of molecules. In addition, the chemical diversity of the different amino acids, together with different chemical environments afforded by local 3D structure, enables many proteins to act as Enzymes, catalyzing a wide range of specific biochemical transformations within cells. In addition, proteins have evolved the ability to bind a wide range of cofactors and Coenzymes, smaller molecules that can endow the protein with specific activities beyond those associated with the polypeptide chain alone.

RNA is Multifunctional

RNA is multifunctional, its primary function is to encode proteins, according to the instructions within a cell's DNA. They control and regulate many aspects of protein synthesis in eukaryotes.

RNA encodes genetic information that can be translated into the amino acid sequence of proteins, as evidenced by the messenger RNA molecules present within every cell, and the RNA genomes of a large number of viruses. The single-stranded nature of RNA, together with tendency for rapid breakdown and a lack of repair systems means that RNA is not so well suited for the long-term storage of genetic information as is DNA.

In addition, RNA is a single-stranded polymer that can, like proteins, fold into a very large number of three-dimensional structures. Some of these structures provide binding sites for other molecules and chemically-active centers that can catalyze specific chemical reactions on those bound molecules. The limited number of different building blocks of RNA (4 nucleotides vs >20 amino acids in proteins), together with their lack of chemical diversity, results in catalytic RNA (ribozymes) being generally less-effective catalysts than proteins for most biological reactions.

Branched Biopolymers

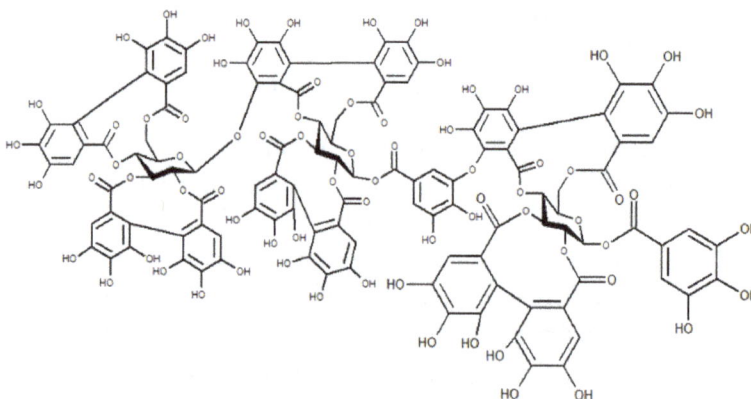

Raspberry ellagitannin, a tannin composed of core of glucose units surrounded
by gallic acid esters and ellagic acid units

Carbohydrate macromolecules (polysaccharides) are formed from polymers of monosaccharides. Because monosaccharides have multiple functional groups, polysaccharides can form linear polymers (e.g. cellulose) or complex branched structures (e.g. glycogen). Polysaccharides perform numerous roles in living organisms, acting as energy stores (e.g. Starch) and as structural components (e.g.chitin in arthropods and fungi). Many carbohydrates contain modified monosaccharide units that have had functional groups replaced or removed.

Polyphenols consist of a branched structure of multiple phenolic subunits. They can perform structural roles (e.g. lignin) as well as roles as secondary metabolites involved in signalling, pigmentation and defense.

Synthetic Macromolecules

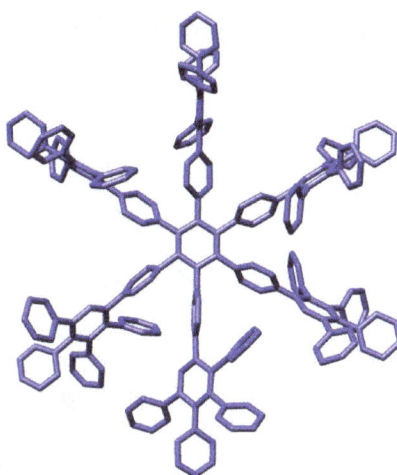

Structure of a polyphenylene dendrimer macromolecule reported by Müllen, et al.

Some examples of macromolecules are synthetic polymers (plastics, synthetic fibers, and synthetic rubber), graphene, and carbon nanotubes. Polymers may be prepared from inorganic matter as well as for instance in inorganic polymers and geopolymers. The incorporation of inorganic elements enables the tunability of properties and/or responsive behavior as for instance in smart inorganic polymers.

Step-growth Polymerization

Step-growth polymerization refers to a type of polymerization mechanism in which bi-functional or multifunctional monomers react to form first dimers, then trimers, longer oligomers and eventually long chain polymers. Many naturally occurring and some synthetic polymers are produced by step-growth polymerization, e.g. polyesters, polyamides, polyurethanes, etc. Due to the nature of the polymerization mechanism, a high extent of reaction is required to achieve high molecular weight. The easiest way to visualize the mechanism of a step-growth polymerization is a group of people reaching out to hold their hands to form a human chain — each person has two hands (= reactive sites). There also is the possibility to have more than two reactive sites on a monomer: In this case branched polymers are produced.

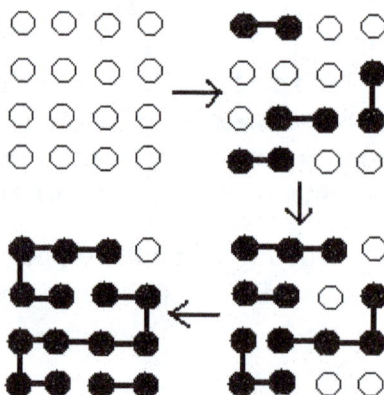

A generic representation of a step-growth polymerization. (Single white dots represent monomers and black chains represent oligomers and polymers)

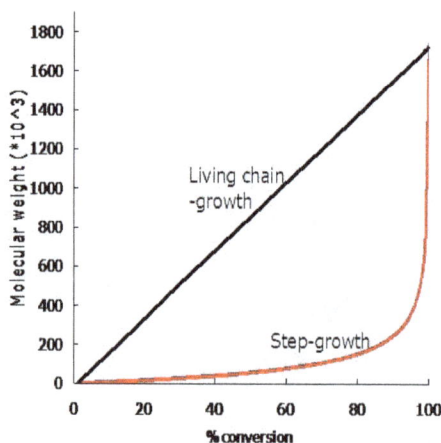

Comparison of Molecular weight vs conversion plot between step-growth and living chain-growth polymerization

Step Growth Polymerization and Condensation Polymerization

"Step growth polymerization" and condensation polymerization are two different concepts, not always identical. In fact polyurethane polymerizes with addition polymerization (because its polymerization produces no small molecules), but its reaction mechanism corresponds to a step-growth polymerization.

The distinction between "addition polymerization" and "condensation polymerization" was introduced by Wallace Hume Carothers in 1929, and refers to the type of products, respectively:

- a polymer only (addition)

- a polymer and a molecule with a low molecular weight (condensation)

The distinction between "step-growth polymerization" and "chain-growth polymerization" was introduced by Paul Flory in 1953, and refers to the reaction mechanisms, respectively:

- by functional groups (step-growth polymerization)

- by free-radical or ion (chain-growth polymerization)

Branched Polymers

A monomer with functionality of 3 or more will introduce branching in a polymer and will ultimately form a cross-linked macrostructure or network even at low fractional conversion. The point at which a tree-like topology transits to a network is known as the gel point because it is signalled by an abrupt change in viscosity. One of the earliest so-called thermosets is known as bakelite. It is not always water that is released in step-growth polymerization: in acyclic diene metathesis or ADMET dienes polymerize with loss of ethylene.

Polymers as Complex Networks

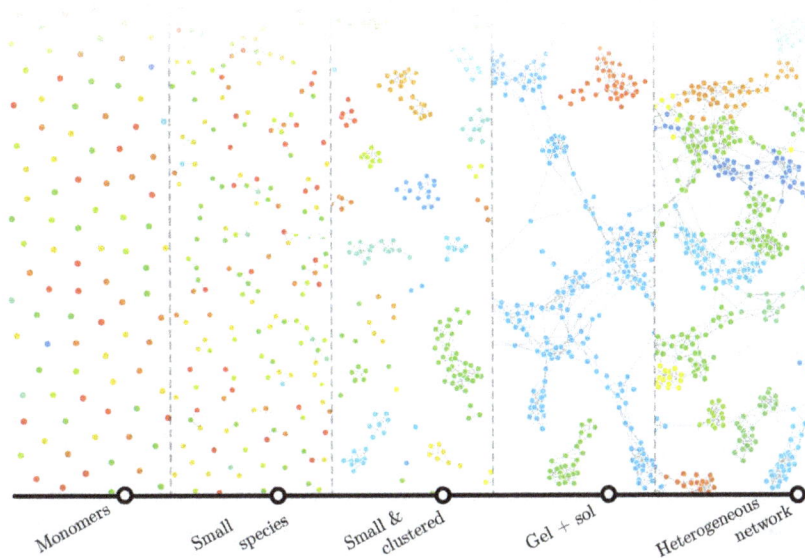

Formation of a polymer network from multifunctional precursors modelled with a random-graph process.

Polymers are macromolecules composed of many repeating units often arranged into complicated networks. In the liquid phase, in a melt or in solution, where each molecule constantly changes its shape as driven by the Brownian motion and advection fields, it is predominately the topology (i.e. the way monomers are arranged into a network) that determines the behaviour of the matter. In most of the cases topology has no regular patterns nor can topologies be observed directly, so one has to rely on models that are meant to predict the formation of a polymer network from scratch by mimicking the particular process of self-assembly or polymerisation. In this respect polymerization is a complex system that can be studied via the lens of complex-network formalism. In this case formation of a molecular network from multifunctional precursors is represented by a random graph process. The process does not account for spatial positions of the monomers explicitly, yet the Euclidean distances between the monomers may derived from the topological information by applying self-avoiding random walks. This allows favoring reactivity of monomers that are close to each other, and to disfavour the reactivity for monomers obscured by the surrounding. The phenomena of conversion-dependent reaction rates, gelation, microgelation, and structural inhomogeneity are predicted by such a model. Moreover, many polymer properties can be extracted from graph theoretical description, these include such descriptors as size distribution, crosslink distances, and gel-point conversion.

Differences between step-growth Polymerization and Chain-growth Polymerization

This technique is usually compared with chain-growth polymerization to show its characteristics.

Step-growth polymerization	Chain-growth polymerization
Growth throughout matrix	Growth by addition of monomer only at one end or both ends of chain
Rapid loss of monomer early in the reaction	Some monomer remains even at long reaction times
Similar steps repeated throughout reaction process	Different steps operate at different stages of mechanism (i.e. Initiation, propagation, termination, and chain transfer)
Average molecular weight increases slowly at low conversion and high extents of reaction are required to obtain high chain length	Molar mass of backbone chain increases rapidly at early stage and remains approximately the same throughout the polymerization
Ends remain active (no termination)	Chains not active after termination
No initiator necessary	Initiator required

Historical Aspects

Most natural polymers being employed at early stage of human society are of condensation type. The synthesis of first truly synthetic polymeric material, Bakelite, was announced by Leo Baekeland in 1907, through a typical step-growth polymerization fashion of phenol and formaldehyde. The pioneer of synthetic polymer science, Wallace H. Carothers, developed a new means of making polyesters through step-growth polymerization in 1930s as a research group leader at DuPont. It was the first reaction designed and carried out with the specific purpose of creating high molecular weight polymer molecules, as well as the first polymerization reaction whose results had been predicted beforehand by scientific theory. Carothers developed a series of mathematic equations to describe the behavior of step-growth polymerization systems which are still known as the Carothers equations today. Collaborating with Paul J. Flory, who is a physical chemist, they developed theories that describe more mathematical aspects of step-growth polymerization including kinetics, stoichiometry, and molecular weight distribution etc. Carothers is also well known for his invention of Nylon.

Classes of Step-growth Polymers

Classes of step-growth polymers are:

- Polyester has high glass transition temperature T_g and high melting point T_m, good mechanical properties to about 175 °C, good resistance to solvent and chemicals. It can exist as fibers and films. The former is used in garments, felts, tire cords, etc. The latter appears in magnetic recording tape and high grade films.

- Polyamide (nylon) has good balance of properties: high strength, good elasticity and abrasion resistance, good toughness, favorable solvent resistance. The applications of polyamide include: rope, belting, fiber cloths, thread, substitute for metal in bearings, jackets on electrical wire.

- Polyurethane can exist as elastomers with good abrasion resistance, hardness, good resistance to grease and good elasticity, as fibers with excellent rebound, as coatings with good resistance to solvent attack and abrasion and as foams with good strength, good rebound and high impact strength.

- Polyurea shows high T_g, fair resistance to greases, oils, and solvents. It can be used in truck bed liners, bridge coating, caulk and decorative designs.

- Polysiloxane are available in a wide range of physical states—from liquids to greases, waxes, resins, and rubbers. Uses of this material are as antifoam and release agents, gaskets, seals, cable and wire insulation, hot liquids and gas conduits, etc.

- Polycarbonates are transparent, self-extinguishing materials. They possess properties like crystalline thermoplasticity, high impact strength, good thermal and oxidative stability. They can be used in machinery, auto-industry, and medical applications. For example, the cockpit canopy of F-22 Raptor is made of high optical quality polycarbonate.

- Polysulfides have outstanding oil and solvent resistance, good gas impermeability, good resistance to aging and ozone. However, it smells bad, and it shows low tensile strength as well as poor heat resistance. It can be used in gasoline hoses, gaskets and places that require solvent resistance and gas resistance.

- Polyether shows good thermoplastic behavior, water solubility, generally good mechanical properties, moderate strength and stiffness. It is applied in sizing for cotton and synthetic fibers, stabilizers for adhesives, binders, and film formers in pharmaceuticals.

- Phenol formaldehyde resin (Bakelite) have good heat resistance, dimensional stability as well as good resistance to most solvents. It also shows good dielectric properties. This material is typically used in molding applications, electrical, radio, televisions and automotive parts where their good dielectric properties are of use. Some other uses include: impregnating paper, varnishes, decorative laminates for wall coverings.

- Poly-Triazole polymers are produced from monomers which bear both an alkyne and azide functional group. The monomer units are linked to each other by the a 1,2,3-triazole group; which is produced by the 1,3-Dipolar cycloaddition, also called the Azide-alkyne Huisgen cycloaddition. These polymers can take on the form of a strong resin, or a gel. With oligopeptide monomers containing a terminal alkyne and terminal azide the resulting clicked peptide polymer will be biodegradable due to action of endopeptidases on the oligopeptide unit.

Kinetics

The kinetics and rates of step-growth polymerization can be described using a polyesterification mechanism. The simple esterification is an acid-catalyzed process in which protonation of the acid is followed by interaction with the alcohol to produce an ester and water. However, there are a few

assumptions needed with this kinetic model. The first assumption is water (or any other condensation product) is efficiently removed. Secondly, the functional group reactivities are independent of chain length. Finally, it is assumed that each step only involves one alcohol and one acid.

$$\frac{1}{1-p^{n-1}} = 1 + (n-1)kt[COOH]^{n-1}$$

This is a general rate law degree of polymerization for polyesterification where n= reaction order.

Self-catalyzed Polyesterification

If no acid catalyst is added, the reaction will still proceed because the acid can act as its own catalyst. The rate of condensation at any time t can then be derived from the rate of disappearance of -COOH groups and

$$rate = \frac{-d[COOH]}{dt} = k[COOH]^2[OH]$$

The second-order [\ce{COOH}] term arises from its use as a catalyst, and k is the rate constant. For a system with equivalent quantities of acid and glycol, the functional group concentration can be written simply as

$$rate = \frac{-d[COOH]}{dt} = k[COOH]^3$$

After integration and substitution from Carothers equation, the final form is the following

$$\frac{1}{(1-p)^2} = 2kt[COOH]^2 + 1 = X_n^2$$

For a self-catalyzed system, the number average degree of polymerization (Xn) grows proportionally with $\sqrt{t}$.

External Catalyzed Polyesterification

The uncatalyzed reaction is rather slow, and a high X_n is not readily attained. In the presence of a catalyst, there is an acceleration of the rate, and the kinetic expression is altered to

$$\frac{-d[COOH]}{dt} = k[COOH][OH]$$

which is kinetically first order in each functional group. Hence,

$$\frac{d[COOH]}{dt} \quad k[COOH]$$

and integration gives finally

$$\frac{1}{1-p} = 1 + [COOH]kt = X_n$$

For an externally catalyzed system, the number average degree of polymerization grows proportionally with t.

Molecular Weight Distribution in Linear Polymerization

The product of a polymerization is a mixture of polymer molecules of different molecular weights. For theoretical and practical reasons it is of interest to discuss the distribution of molecular weights in a polymerization. The molecular weight distribution (MWD) had been derived by Flory by a statistical approach based on the concept of equal reactivity of functional groups.

Probability

Step-growth polymerization is a random process so we can use statistics to calculate the probability of finding a chain with x-structural units ("x-mer") as a function of time or conversion.

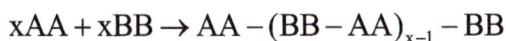

$$xAA + xBB \rightarrow AA - (BB - AA)_{x-1} - BB$$

$$xAB \rightarrow A - (B - A)_{x-1} - B$$

Probability that an 'A' functional group has reacted

$$p^{x-1}$$

Probability of finding an 'A' unreacted

$$(1 - p)$$

Combining the above two equations leads to.

$$P_x = (1 - p)p^{x-1}$$

Where P_x is the probability of finding a chain that is x-units long and has an unreacted 'A'. As x increases the probability decreases.

Number Fraction Distribution

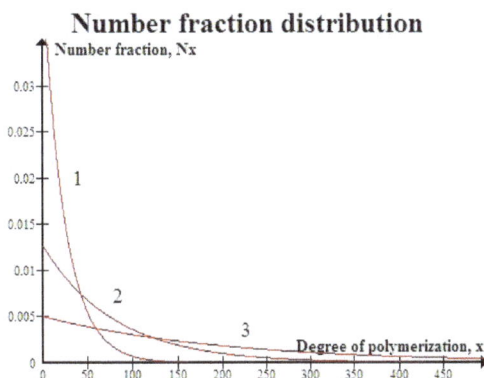

Number-fraction distribution curve for linear polymerization. Plot 1, p=0.9600; plot 2, p=0.9875; plot 3, p=0.9950.

The number fraction distribution is the fraction of x-mers in any system and equals the probability of finding it in solution.

$$\frac{N_x}{N} = (1 - p)p^{x-1}$$

Where N is the total number of polymer molecules present in the reaction.

Weight Fraction Distribution

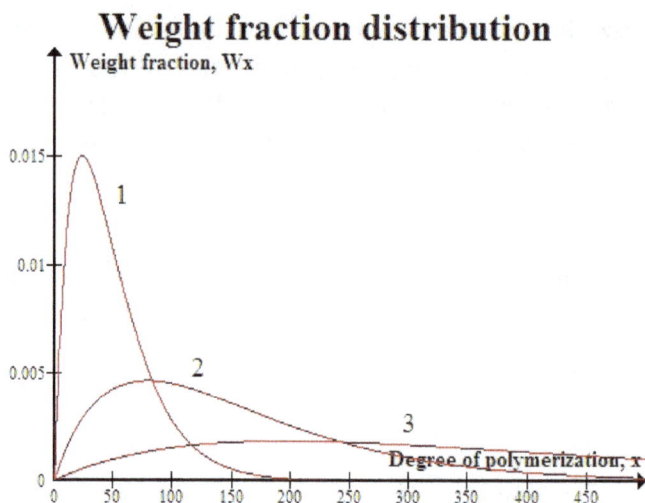

Weight fraction distribution plot for linear polymerization. Plot 1, p=0.9600; plot 2, p=0.9875; plot 3, p=0.9950.

The weight fraction distribution is the fraction of x-mers in a system and the probability of finding them in terms of mass fraction.

$$\frac{W_x}{W_o} = \frac{xN_xM_o}{N_oM_o} = \frac{xN_x}{N_o} = x\frac{N_x}{N}\frac{N}{N_o}$$

Notes:

- M_o is the molar mass of the repeat unit,

- N_o is the initial number of monomer molecules,

- and N is the number of unreacted functional groups

Substituting from the Carothers equation

$$X_n = \frac{1}{1-p} = \frac{N_o}{N}$$

We can now obtain:

$$\frac{W_x}{W_o} = x(1-p)^2 p^{x-1}$$

PDI

The polydispersity index (PDI), is a measure of the distribution of molecular mass in a given polymer sample.

$$PDI = \frac{M_w}{M_n}$$

However for step-growth polymerization the Carothers equation can be used to substitute and rearrange this formula into the following.

$$PDI = 1 + p$$

Therefore, in step-growth when p=1, then the PDI=2.

Molecular Weight Control in Linear Polymerization

Need for Stoichiometric Control

There are two important aspects with regard to the control of molecular weight in polymerization. In the synthesis of polymers, one is usually interested in obtaining a product of very specific molecular weight, since the properties of the polymer will usually be highly dependent on molecular weight. Molecular weights higher or lower than the desired weight are equally undesirable. Since the degree of polymerization is a function of reaction time, the desired molecular weight can be obtained by quenching the reaction at the appropriate time. However, the polymer obtained in this manner is unstable in that it leads to changes in molecular weight because the ends of the polymer molecule contain functional groups that can react further with each other.

This situation is avoided by adjusting the concentrations of the two monomers so that they are slightly nonstoichiometric. One of the reactants is present in slight excess. The polymerization then proceeds to a point at which one reactant is completely used up and all the chain ends possess the same functional group of the group that is in excess. Further polymerization is not possible, and the polymer is stable to subsequent molecular weight changes.

Another method of achieving the desired molecular weight is by addition of a small amount of monofunctional monomer, a monomer with only one functional group. The monofunctional monomer, often referred to as a chain stopper, controls and limits the polymerization of bifunctional monomers because the growing polymer yields chain ends devoid of functional groups and therefore incapable of further reaction.

Quantitative Aspects

To properly control the polymer molecular weight, the stoichiometric imbalance of the bifunctional monomer or the monofunctional monomer must be precisely adjusted. If the nonstoichiometric imbalance is too large, the polymer molecular weight will be too low. It is important to understand the quantitative effect of the stoichiometric imbalance of reactants on the molecular weight. Also, this is necessary in order to know the quantitative effect of any reactive impurities that may be present in the reaction mixture either initially or that are formed by undesirable side reactions. Impurities with A or B functional groups may drastically lower the polymer molecular weight unless their presence is quantitatively taken into account.

More usefully, a precisely controlled stoichiometric imbalance of the reactants in the mixture can provide the desired result. For example, an excess of diamine over an acid chloride would eventually produce a polyamide with two amine end groups incapable of further growth when the acid chloride was totally consumed. This can be expressed in an extension of the Carothers equation as,

$$X_n = \frac{(1+r)}{(1+r-2rp)}$$

were r is the ratio of the number of molecules of the reactants.

$$r = \frac{N_{AA}}{N_{BB}}$$ were N_{BB} is the molecule in excess.

The equation above can also be used for a monofunctional additive which is the following,

$$r = \frac{N_{AA}}{(N_{BB} + 2N_B)}$$

where N_B is the number of monofunction molecules added. The coefficient of 2 in front of N_B is require since one B molecule has the same quantitative effect as one excess B-B molecule.

Multi-chain Polymerization

A monomer with functionality 3 has 3 functional groups which participate in the polymerization. This will introduce branching in a polymer and may ultimately form a cross-linked macrostructure. The point at which this three-dimensional network is formed is known as the gel point, signaled by an abrupt change in viscosity.

A more general functionality factor f_{av} is defined for multi-chain polymerization, as the average number of functional groups present per monomer unit. For a system containing N_0 molecules initially and equivalent numbers of two function groups A and B, the total number of functional groups is $N_0 f_{av}$.

$$f_{av} = \frac{\sum N_i \cdot f_i}{\sum N_i}$$

And the modified Carothers equation is

$$x_n = \frac{2}{2 - pf_{av}},$$ where p equals to $$\frac{2(N_0 - N)}{N_0 \cdot f_{av}}$$

Advances in step-growth Polymers

The driving force in designing new polymers is the prospect of replacing other materials of construction, especially metals, by using lightweight and heat-resistant polymers. The advantages of lightweight polymers include: high strength, solvent and chemical resistance, contributing to a variety of potential uses, such as electrical and engine parts on automotive and aircraft components, coatings on cookware, coating and circuit boards for electronic and microelectronic devices, etc. Polymer chains based on aromatic rings are desirable due to high bond strengths and rigid polymer chains. High molecular weight and crosslinking are desirable for the same reason. Strong dipole-dipole, hydrogen bond interactions and crystallinity also improve heat resistance. To obtain desired mechanical strength, sufficiently high molecular weights are necessary, however, decreased solubility is a problem. One approach to solve this problem is to introduce of some flexibilizing linkages, such as isopropylidene, C=O, and SO 2 into the rigid polymer chain by using an appropriate monomer or comonomer. Another approach involves the synthesis of reactive telechelic oligomers containing functional end groups capable of reacting with each other, polymerization of the oligomer gives higher molecular weight, referred to as chain extension.

Aromatic Polyether

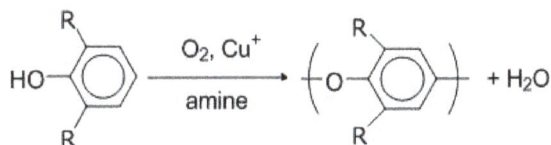

The oxidative coupling polymerization of many 2,6-disubstituted phenols using a catalytic complex of a cuprous salt and amine form aromatic polyethers, commercially referred to as poly(p-phenylene oxide) or PPO. Neat PPO has little commercial uses due to its high melt viscosity. Its available products are blends of PPO with high-impact polystyrene (HIPS).

Polyethersulfone

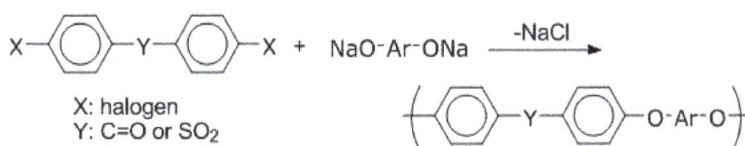

X: halogen
Y: C=O or SO$_2$

Polyethersulfone (PES) is also referred to as polyetherketone, polysulfone. It is synthesized by nucleophilic aromatic substitution between aromatic dihalides and bisphenolate salts. Polyethersulfones are partially crystalline, highly resistant to a wide range of aqueous and organic environment. They are rated for continuous service at temperatures of 240-280 °C. The polyketones are finding applications in areas like automotive, aerospace, electrical-electronic cable insulation.

Aromatic Polysulfides

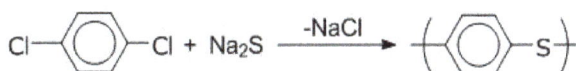

Poly(p-phenylene sulfide) (PPS) is synthesized by the reaction of sodium sulfide with p-dichlorobenzene in a polar solvent such as 1-methyl-2-pyrrolidinone (NMP). It is inherently flame-resistant and stable toward organic and aqueous conditions; however, it is somewhat susceptible to oxidants. Applications of PPS include automotive, microwave oven component, coating for cookware when blend with fluorocarbon polymers and protective coatings for valves, pipes, electromotive cells, etc.

Aromatic Polyimide

Aromatic polyimides are synthesized by the reaction of dianhydrides with diamines, for example, pyromellitic anhydride with p-phenylenediamine. It can also be accomplished using diisocyanates in place of diamines. Solubility considerations sometimes suggest use of the half acid-half ester of the dianhydride, instead of the dianhydride itself. Polymerization is accomplished by a two-stage process due to the insolubility of polyimides. The first stage forms a soluble and fusible high-molecular-weight poly(amic acid) in a polar aprotic solvent such as NMP or N,N-dimethylacetamide. The poly(amic aicd) can then be processed into the desired physical form of the final plymer product (e.g., film, fiber, laminate, coating) which is insoluble and infusible.

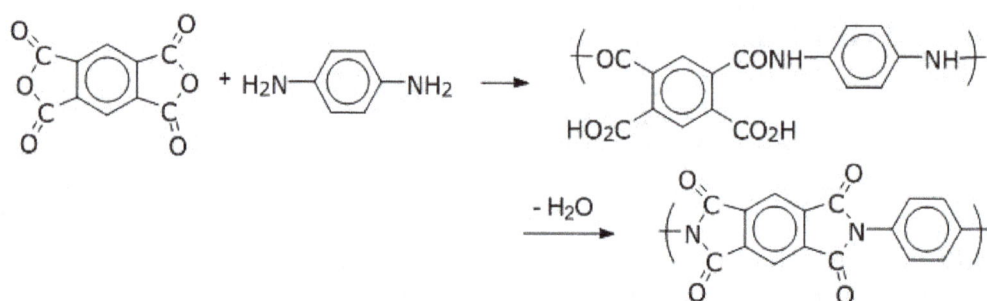

Telechelic Oligomer Approach

Telechelic oligomer approach applies the usual polymerization manner except that one includes a monofunctional reactant to stop reaction at the oligomer stage, generally in the 50-3000 molecular weight. The monofunctional reactant not only limits polymerization but end-caps the oligomer with functional groups capable of subsequent reaction to achieve curing of the oligomer. Functional groups like alkyne, norbornene, maleimide, nitrite, and cyanate have been used for this purpose. Maleimide and norbornene end-capped oligomers can be cured by heating. Alkyne, nitrile, and cyanate end-capped oligomers can undergo cyclotrimerization yielding aromatic structures.

Chain-growth Polymerization

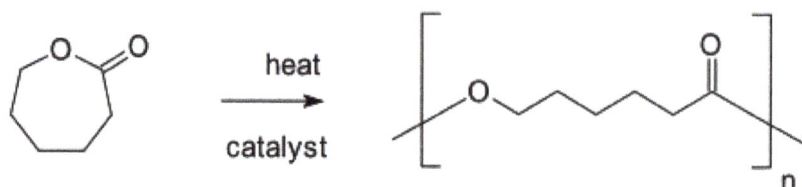

An example of chain-growth polymerization by ring opening to polycaprolactone

Chain-growth polymerization or chain polymerization is a polymerization technique where unsaturated monomer molecules add onto the active site of a growing polymer chain one at a time. Growth of the polymer occurs only at one (or possibly more) ends. Addition of each monomer unit regenerates the active site.

Polyethylene, polypropylene, and polyvinyl chloride (PVC) are common types of plastics made by chain-growth polymerization. They are the primary component of four of the plastics specifically labeled with recycling codes and are used extensively in packaging.

Mechanism

Chain-growth polymerization can be understood with the chemical equation:

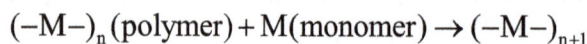

$$(-M-)_n (polymer) + M (monomer) \rightarrow (-M-)_{n+1}$$

where n is the degree of polymerization and M is some form of unsaturated compound: an alkene (vinyl polymers) or alicyclic compound (ring-opening polymerization) containing molecule.

This type of polymerization result in high molecular weight polymer being formed at low conversion. This final weight is determined by the rate of propagation compared to the rate of individual chain termination, which includes both chain transfer and chain termination steps. Above a certain ceiling temperature, no polymerization occurs.

Steps

Chain-growth polymerization usually has the following steps:

1. chain initiation, usually by means of an initiator which starts the chemical process. Typical initiators include any organic compound with a labile group: e.g. azo (-N=N-), disulfide (-S-S-), or peroxide (-O-O-). Two examples are benzoyl peroxide and AIBN.

2. chain propagation

3. chain transfer, terminates the chain, but the active site is transferred to a new chain. This can occur with the solvent, monomer, or other polymer. This process increases the branching of the resulting polymer.

4. chain termination, which occurs either by combination or disproportionation. Termination, in radical polymerization, is when the free radicals combine and is the end of the polymerization process.

The active center can be one of a number of different types:

- free radical in radical polymerization, for example, polystyrene, sometimes seen as packing peanuts, is produced by polymerizing styrene with Benzoyl peroxide as its radical initiator

- carbocation in cationic polymerization, an example is Isobutyl synthetic rubber, initiated by Aluminium chloride ionizing isobutylene

- carbanion in anionic polymerization

- organometallic complex in coordination polymerization.

Under the necessary reaction conditions, an addition polymerization can be considered a living polymerization. This is most often seen with anionic polymerization as it can be easy to perform without termination steps.

Comparison with other Polymerization Methods

The distinction between step-growth polymerization and chain-growth polymerization was introduced by Paul Flory in 1953, and refers to the difference in reaction mechanisms with step-growth using the functional groups of the monomer compared to the free-radical or ion groups used in chain-growth polymerization.

Chain growth polymerization and addition polymerization (also called polyaddition) are two different concepts. In fact polyurethane polymerizes with addition polymerization (because its polymerization does not produce any small molecules, called "condensate"), but its reaction mechanism is a step-growth polymerization.

The distinction between "addition polymerization" and "condensation polymerization" was introduced by Wallace Hume Carothers in 1929, and refers to the type of product produced. Addition polymerization produces only a polymer molecule, while condensation polymerization produces a polymer as well as a molecule with a low molecular weight, usually water.

Forms of Chain-growth Polymerization

Anionic Addition Polymerization

Anionic addition polymerization is a form of chain-growth polymerization or addition polymerization that involves the polymerization of vinyl monomers with strong electronegative groups. This polymerization is carried out through a carbanion active species. Like all chain-growth polymerizations, it takes place in three steps: chain initiation, chain propagation, and chain termination. Living polymerizations, which lack a formal termination pathway, occur in many anionic addition polymerizations. The advantage of living anionic addition polymerizations is that they allow for the control of structure and composition.

Anionic polymerizations are used in the production of polydiene synthetic rubbers, solution styrene/butadiene rubbers (SBR), and styrenic thermoplastic elastomers.

History

As early as 1936, Karl Ziegler proposed that anionic polymerization of styrene and butadiene by consecutive addition of monomer to an alkyl lithium initiator occurred without chain transfer or termination. Twenty years later, living polymerization was demonstrated by Szwarc. The early work of Michael Szwarc and co – workers in 1956 was one of the breakthrough events in the field of polymer science. When Szwarc learned that the electron transfer between radical anion of naphthalene and styrene in an aprotic solvent such as tetrahydrofuran gave a messy product, he started investigating the reaction in more detail. He proved that the electron transfer results in the formation of a dianion which rapidly added styrene to form a "two – ended living polymer." Being a physical chemist, Szwarc set forth in understanding the mechanism of such living polymerization in greater detail. His work elucidated the kinetics and the thermodynamics of the process in considerable detail. At the same time, he explored the structure property relationship of the various ion pairs and radical ions involved. This had great ramifications in future research in polymer synthesis, because Szwarc had found a way to make polymers with greater control over molecular weight, molecular weight distribution and the architecture of the polymer.

The use of alkali metals to initiate polymerization of 1,3-dienes led to the discovery by Stavely and co-workers at Firestone Tire and Rubber company of cis-1,4-polyisoprene. This sparked the development of commercial anionic polymerization processes that utilize alkyllithium initiatiors.

Monomer Characteristics

In order for polymerization to occur with vinyl monomers, the substituents on the double bond must be able to stabilize a negative charge. Stabilization occurs through delocalization of the negative charge. Because of the nature of the carbanion propagating center, substituents that react with bases or nucleophiles either must not be present or be protected.

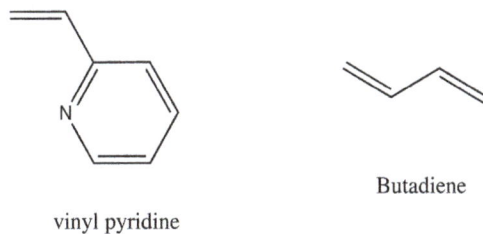

vinyl pyridine

Butadiene

Examples of vinyl monomers.

Vinyl monomers with substituents that stabilize the negative charge through charge delocalization, undergo polymerization without termination or chain transfer. These monomers include styrene, dienes, methacrylate, vinyl pyridine, aldehydes, epoxide, episulfide, cyclic siloxane, and lactones. Polar monomers, using controlled conditions and low temperatures, can undergo anionic polymerization. However, at higher temperatures they do not produce living stable, carbanionic chain ends because their polar substituents can undergo side reactions with both initiators and propagating chain centers. The effects of counterion, solvent, temperature, Lewis base additives, and inorganic solvents have been investigated to increase the potential of anionic polymerizations of polar monomers. Polar monomers include acrylonitrile, cyanoacrylate, propylene oxide, vinyl ketone, acrolein, vinyl sulfone, vinyl sulfoxide, vinyl silane and isocyanate.

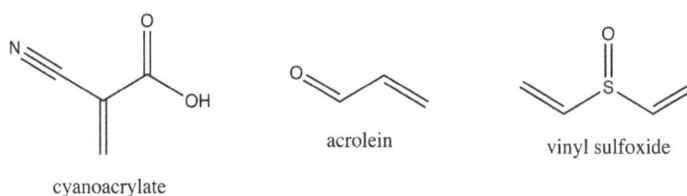

cyanoacrylate

acrolein

vinyl sulfoxide

Examples of polar monomers.

Solvent

The solvent used in anionic addition polymerizations are determined by the reactivity of both the initiator and carbanion of the propagating chain end. Anionic species with low reactivity, such as heterocyclic monomers, can use a wide range of solvents.

Initiation

The reactivity of initiators used in anionic polymerization should be similar to that of the monomer that is the propagating species. The pKa values for the conjugate acids of the carbanions formed from monomers can be used to deduce the reactivity of the monomer. The least reactive monomers have the largest pKa values for their corresponding conjugate acid and thus, require the most reactive initiator. Two main initiation pathways involve electron transfer (through alkali metals) and strong anions.

Initiation by Electron Transfer

Szwarc and coworkers studied the initiation of polymerization through the use of aromatic radical-anions such as sodium naphthenate. In this reaction, an electron is transferred from the alkali metal to naphthalene. Polar solvents are necessary for this type of initiation both for stability of the anion-radical and to solvate the cation species formed. The anion-radical can then transfer an electron to the monomer.

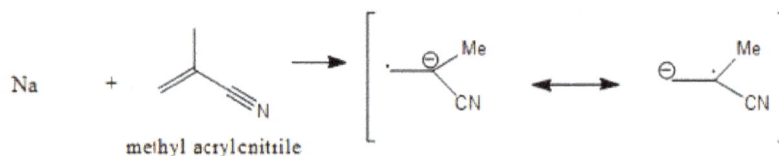

Initiation through electron transfer.

Initiation can also involve the transfer of an electron from the alkali metal to the monomer to form an anion-radical. Initiation occurs on the surface of the metal, with the reversible transfer of an electron to the adsorbed monomer.

Initiation by Strong Anions

Nucleophilic initiators include covalent or ionic metal amides, alkoxides, hydroxides, cyanides, phosphines, amines and organometallic compounds (alkyllithium compounds and Grignard reagents). The initiation process involves the addition of a neutral (B:) or negative (B:-) nucleophile to the monomer.

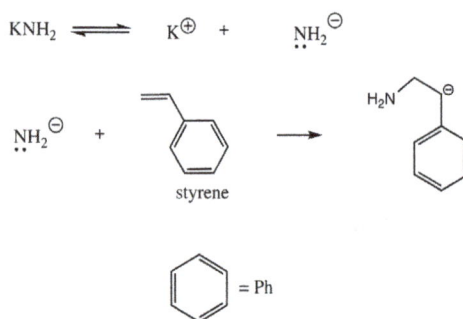

Initiation through strong anion.

The most commercially useful of these initiators has been the alkyllithium initiators. They are primarily used for the polymerization of styrenes and dienes.

Monomers activated by strong electronegative groups may be initiated even by weak anionic or neutral nucleophiles (i.e. amines, phosphines). Most prominent example is the curing of cyanoacrylate, which constitutes the basis for superglue. Here, only traces of basic impurities are sufficient to induce an anionic addition polymerization or zwitterionic addition polymerization, respectively.

Propagation

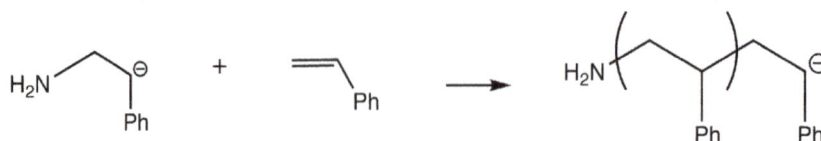

Propagation of an anionic addition polymerization.

Propagation in anionic addition polymerization results in the complete consumption of monomer. It is very fast and occurs at low temperatures. This is due to the anion not being very stable, the speed of the reaction as well as that heat is released during the reaction. The stability can be greatly enhanced by reducing the temperatures to near $0\,^{\circ}C$. The propagation rates are generally fairly high compared to the decay reaction, so the overall polymerization rates is generally not affected.

Termination

Anionic addition polymerizations have no formal termination pathways because proton transfer from solvent or other positive species does not occur. However, termination can occur through unintentional quenching due to trace impurities. This includes trace amounts of oxygen, carbon dioxide or water. Intentional termination can occur through the addition of water or alcohol. Another method of termination, chain transfer, can occur when an agent can act as a Brønsted acid. In this case, the pKa value of the agent is similar to the conjugate acid of the propagating carbanionic chain end. Spontaneous termination occurs because the concentration of carbanion centers decay over time and eventually results in hydride elimination. Polar monomers are more reactive because they are stabilized by their polar substituents. These polar substituents can react with nucleophiles which results in termination as well as side reactions that compete with both initiation and propagation.

Living Anionic Polymerization

Living polymerization was demonstrated by Szwarc and co workers in 1956. Their initial work was based on the polymerization of styrene and dienes. One of the remarkable features of living anionic polymerization is that the mechanism involves no formal termination step. In the absence of impurities, the carbanion would still be active and capable of adding another monomer. The chains will remain active indefinitely unless there is inadvertent or deliberate termination or chain transfer.

Kinetics

The kinetics of anionic addition polymerization depend on whether or not a termination pathway occurs.

Kinetics of Living Anionic Addition Polymerization

In general, the reaction mechanism for living anionic addition polymerization are as follows:

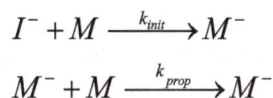

$$I^- + M \xrightarrow{k_{init}} M^-$$

$$M^- + M \xrightarrow{k_{prop}} M^-$$

where I = initiator, k_{init} = the initiation reaction rate constant, M = monomer, M⁻= propagating species, and k_{prop} = the propagation reaction rate constant.

As most polymerizations of this type do not have a termination pathway, the rate of polymerization is the rate of propagation:

$$\text{rate(prop)} = k_p[M^-][M]$$

where k_p is the rate of constant of propagation, $[M^-]$ is the total concentration of propagating centers, and $[M]$ is the concentration of monomer. Since there is no termination pathway in living polymerizations, the concentration of propagating centers is equal to the concentration of initiator ($[I]$). Thus,

$$\text{rate(prop)} = k_p[I][M]$$

The degree of polymerization, X_n is also affected by no termination pathway. It is the ratio of concentration of reacted monomer ($[M]_o$) to initiator($[I]_o$) times the percent conversion p. In this case, the chain length (v) is equal to X_n.

$$v = \frac{[M]_o}{[I]_o} \rho$$

When conversion, $p = 1$ (100% conversion), chain length is simply the ratio of reacted monomer to initiator.

$$v = \frac{[M]_o}{[I]_o}$$

Kinetics: Termination due to Impurities

When termination occurs due to impurities, the impurities must be taken into account in determining the reaction rate. The reaction mechanisms would begin the same as that of a living anionic addition (initiation and propagation). However, there would now be a termination step to account for the effect of the impurities on the reaction.

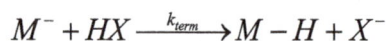

$$M^- + HX \xrightarrow{k_{term}} M - H + X^-$$

where M^- = propagating species, HX = impurity and k_{term} = the termination reaction rate constant.

Using the steady-state approximation, the rate of propagation becomes

$$\text{rate(prop)} = \frac{k_{init}k_{prop}[I][M]^2}{k_{term}[H-X]}$$

Since

$$v = \frac{\text{rate(prop)}}{\text{rate(term)}} = \frac{k_{prop}[M]}{k_{term}[H-X]}$$

Thus chain length and rate of propagation are negatively impacted by the presence of impurities in the reaction.

Cationic Polymerization

Cationic polymerization is a type of chain growth polymerization in which a cationic initiator transfers charge to a monomer which then becomes reactive. This reactive monomer goes on to react similarly with other monomers to form a polymer. The types of monomers necessary for cationic polymerization are limited to olefins with electron-donating substituents and heterocycles. Similar to anionic polymerization reactions, cationic polymerization reactions are very sensitive to the type of solvent used. Specifically, the ability of a solvent to form free ions will dictate the reactivity of the propagating cationic chain. Cationic polymerization is used in the production of polyisobutylene (used in inner tubes) and poly(N-vinylcarbazole) (PVK).

IUPAC Definition

An ionic polymerization in which the kinetic-chain carriers are cations.

Monomers

Monomer scope for cationic polymerization is limited to two main types: olefins and heterocyclic monomers. Cationic polymerization of both types of monomers occurs only if the overall reaction is thermally favorable. In the case of olefins, this is due to isomerization of the monomer double bond; for heterocycles, this is due to release of monomer ring strain and, in some cases, isomerization of repeating units. Monomers for cationic polymerization are nucleophilic and form a stable cation upon polymerization.

Olefins

Cationic polymerization of olefin monomers occurs with olefins that contain electron-donating substituents. These electron-donating groups make the olefin nucleophilic enough to attack electrophilic initiators or growing polymer chains. At the same time, these electron-donating groups attached to the monomer must be able to stabilize the resulting cationic charge for further polymerization. Some reactive olefin monomers are shown below in order of decreasing reactivity, with heteroatom groups being more reactive than alkyl or aryl groups. Note, however, that the reactivity of the carbenium ion formed is the opposite of the monomer reactivity.

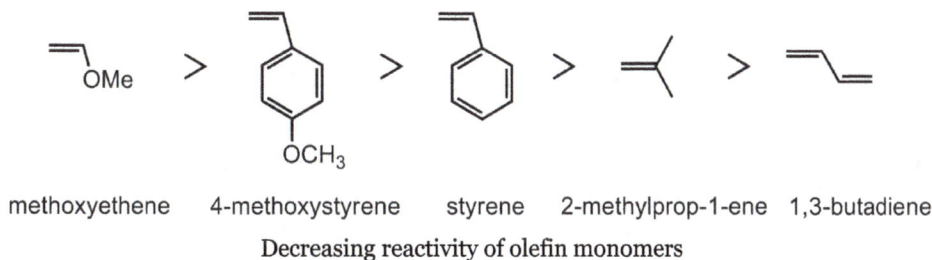

methoxyethene   4-methoxystyrene   styrene   2-methylprop-1-ene   1,3-butadiene
Decreasing reactivity of olefin monomers

Heterocyclic Monomers

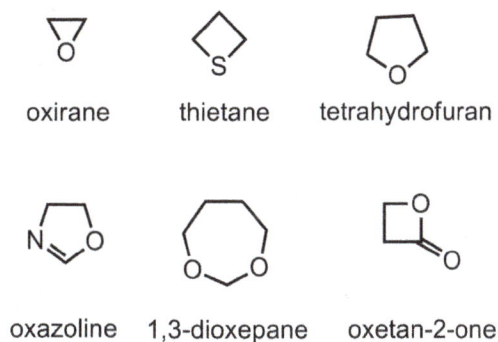

oxirane   thietane   tetrahydrofuran

oxazoline   1,3-dioxepane   oxetan-2-one
Examples of heterocyclic monomers

Heterocyclic monomers that are cationically polymerized are lactones, lactams, and cyclic amines. Upon addition of an initiator, cyclic monomers go on to form linear polymers. The reactivity of heterocyclic monomers depends on their ring strain. Monomers with large ring strain, such as ox-irane, are more reactive than 1,3-dioxepane which has considerably less ring strain. Rings that are six-membered and larger are less likely to polymerize due to lower ring strain.

Synthesis

Initiation

Initiation is the first step in cationic polymerization. During initiation, a carbenium ion is generated from which the polymer chain is made. The counterion should be non-nucleophilic, otherwise the reaction is terminated instantaneously. There are a variety of initiators available for cationic polymerization, and some of them require a coinitiator to generate the needed cationic species.

Classical Protonic Acids

Strong protic acids can be used to form a cationic initiating species. High concentrations of the acid are needed in order to produce sufficient quantities of the cationic species. The counterion (A^-) produced must be weakly nucleophilic so as to prevent early termination due to combination with the protonated olefin. Common acids used are phosphoric, sulfuric, fluro-, and triflic acids. Only low molecular weight polymers are formed with these initiators.

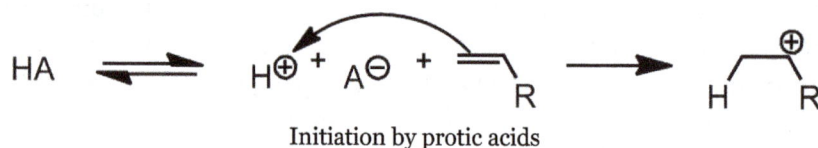

Initiation by protic acids

Lewis Acids/Friedel-Crafts Catalysts

Lewis acids are the most common compounds used for initiation of cationic polymerization. The more popular Lewis acids are $SnCl_4$, $AlCl_3$, BF_3, and $TiCl_4$. Although these Lewis acids alone are able to induce polymerization, the reaction occurs much faster with a suitable cation source. The cation source can be water, alcohols, or even a carbocation donor such as an ester or an anhydride. In these systems the Lewis acid is referred to as a coinitiator while the cation source is the initiator. Upon reaction of the initiator with the coinitiator, an intermediate complex is formed which then goes on to react with the monomer unit. The counterion produced by the initiator-coinitiator complex is less nucleophilic than that of the Brønsted acid A^- counterion. Halogens, such as chlorine and bromine, can also initiate cationic polymerization upon addition of the more active Lewis acids.

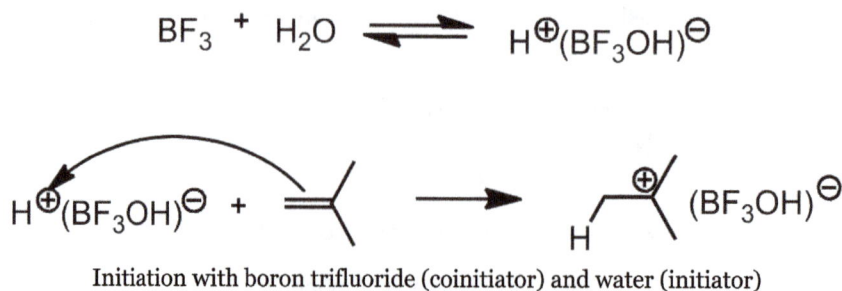

$$BF_3 + H_2O \rightleftharpoons H^{\oplus}(BF_3OH)^{\ominus}$$

Initiation with boron trifluoride (coinitiator) and water (initiator)

Carbenium Ion Salts

Stable carbenium ions are used to initiate chain growth of only the most reactive olefins and are known to give well defined structures. These initiators are most often used in kinetic studies due to the ease of being able to measure the disappearance of the carbenium ion absorbance. Common carbenium ions are trityl and tropylium cations.

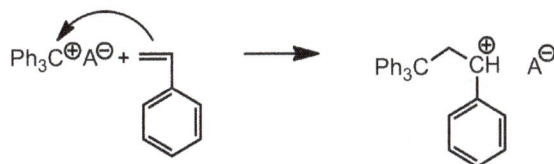

Initiation with trityl carbenium ion

Ionizing Radiation

Ionizing radiation can form a radical-cation pair that can then react with a monomer to start cationic polymerization. Control of the radical-cation pairs are difficult and often depend on the monomer and reaction conditions. Formation of radical and anionic species are often observed.

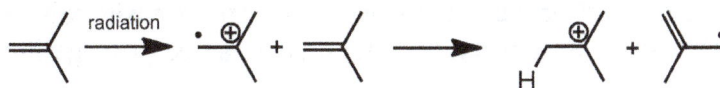

Initiation using ionizing radiation

Propagation

Propagation proceeds via addition of monomer to the active species, i.e. the carbenium ion. The monomer is added to the growing chain in a head-to-tail fashion; in the process, the cationic end group is regenerated to allow for the next round of monomer addition.

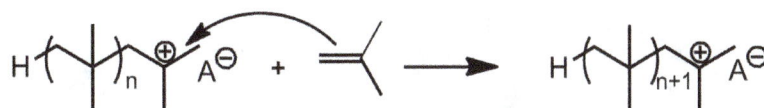

General propagation pathway

Effect of Temperature

The temperature of the reaction has an effect on the rate of propagation. The overall activation energy for the polymerization (E) is based upon the activation energies for the initiation (E_i), propagation (E_p), and termination (v) steps:

$$E = E_i + E_p - E_t$$

Generally, E_t is larger than the sum of E_i and E_p, meaning the overall activation energy is negative. When this is the case, a decrease in temperature leads to an increase in the rate of propagation. The converse is true when the overall activation energy is positive.

Chain length is also affected by temperature. Low reaction temperatures, in the range of 170–190 K, are preferred for producing longer chains. This comes as a result of the activation energy for termination and other side reactions being larger than the activation energy for propagation. As the temperature is raised, the energy barrier for the termination reaction is overcome, causing shorter chains to be produced during the polymerization process.

Effect of Solvent and Counterion

The solvent and the counterion (the gegen ion) have a significant effect on the rate of propagation. The counterion and the carbenium ion can have different associations according to intimate ion pair theory; ranging from a covalent bond, tight ion pair (unseparated), solvent-separated ion pair (partially separated), and free ions (completely dissociated).

The association is strongest as a covalent bond and weakest when the pair exists as free ions. In cationic polymerization, the ions tend to be in equilibrium between an ion pair (either tight or solvent-separated) and free ions. The more polar the solvent used in the reaction, the better the solvation and separation of the ions. Since free ions are more reactive than ion pairs, the rate of propagation is faster in more polar solvents.

The size of the counterion is also a factor. A smaller counterion, with a higher charge density, will have stronger electrostatic interactions with the carbenium ion than will a larger counterion which has a lower charge density. Further, a smaller counterion is more easily solvated by a polar solvent than a counterion with low charge density. The result is increased propagation rate with increased solvating capability of the solvent.

Termination

Termination generally occurs via unimolecular rearrangement with the counterion. In this process, an anionic fragment of the counterion combines with the propagating chain end. This not only inactivates the growing chain, but it also terminates the kinetic chain by reducing the concentration of the initiator-coinitiator complex.

Termination by combination with an anionic fragment from the counterion

Chain Transfer

Chain transfer can take place in two ways. One method of chain transfer is hydrogen abstraction from the active chain end to the counterion. In this process, the growing chain is terminated, but the initiator-coinitiator complex is regenerated to initiate more chains.

Chain transfer by hydrogen abstraction to the counterion

The second method involves hydrogen abstraction from the active chain end to the monomer. This terminates the growing chain and also forms a new active carbenium ion-counterion complex which can continue to propagate, thus keeping the kinetic chain intact.

Chain transfer by hydrogen abstraction to the monomer

Cationic Ring-opening Polymerization

Cationic ring-opening polymerization follows the same mechanistic steps of initiation, propagation, and termination. However, in this polymerization reaction, the monomer units are cyclic in comparison to the resulting polymer chains which are linear. The linear polymers produced can have low ceiling temperatures, hence end-capping of the polymer chains is often necessary to prevent depolymerization.

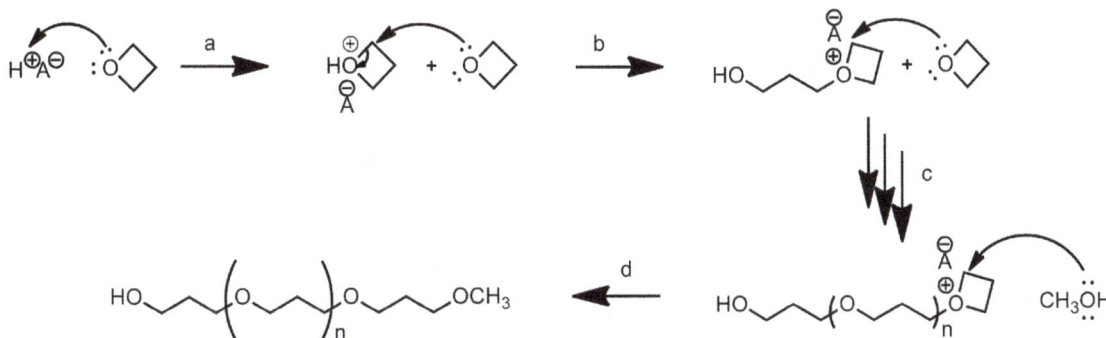

Cationic ring-opening polymerization of oxetane involving (a and b) initiation, (c) propagation, and (d) termination with methanol

Kinetics

The rate of propagation and the degree of polymerization can be determined from an analysis of the kinetics of the polymerization. The reaction equations for initiation, propagation, termination, and chain transfer can be written in a general form:

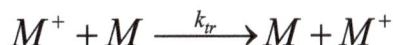

$$I^+ + M \xrightarrow{k_i} M^+$$

$$M^+ + M \xrightarrow{kp} M^+$$

$$M^+ \xrightarrow{k_t} M$$

$$M^+ + M \xrightarrow{k_{tr}} M + M^+$$

In which I^+ is the initiator, M is the monomer, M^+ is the propagating center, and k_i, k_p, k_t, and k_{tr} are the rate constants for initiation, propagation, termination, and chain transfer, respectively. For simplicity, counterions are not shown in the above reaction equations and only chain transfer to monomer is considered. The resulting rate equations are as follows, where brackets denote concentrations:

$$\text{rate(initiation)} = k_i[I^+][M]$$

$$\text{rate(propagation)} = k_p[M^+][M]$$

$$\text{rate(termination)} = k_t[M^+]$$

$$\text{rate(chain transfer)} = k_{tr}[M^+][M]$$

Assuming steady-state conditions, i.e. the rate of initiation = rate of termination:

$$[M^+] = \frac{k_i[I^+][M]}{k_t}$$

This equation for [M+] can then be used in the equation for the rate of propagation:

$$\text{rate(propagation)} = \frac{k_p k_i [M]^2 [I^+]}{k_t}$$

From this equation, it is seen that propagation rate increases with increasing monomer and initiator concentration.

The degree of polymerization, X_n, can be determined from the rates of propagation and termination:

$$X_n = \frac{\text{rate(propagation)}}{\text{rate(termination)}} = \frac{k_p[M]}{k_t}$$

If chain transfer rather than termination is dominant, the equation for X_n becomes

$$X_n = \frac{\text{rate(propagation)}}{\text{rate(chain transfer)}} = \frac{k_p}{k_{tr}}$$

Living Polymerization

In 1984, Higashimura and Sawamoto reported the first living cationic polymerization for alkyl vinyl ethers. This type of polymerization has allowed for the control of well-defined polymers. A key characteristic of living cationic polymerization is that termination is essentially eliminated, thus the cationic chain growth continues until all monomer is consumed.

Commercial Applications

The largest commercial application of cationic polymerization is in the production of polyisobutylene (PIB) products which include polybutene and butyl rubber. These polymers have a variety of applications from adhesives and sealants to protective gloves and pharmaceutical stoppers. The reaction conditions for the synthesis of each type of isobutylene product vary depending on the desired molecular weight and what type(s) of monomer(s) is used. The conditions most commonly used to form low molecular weight ($5-10 \times 10^4$ Da) polyisobutylene are initiation with $AlCl_3$, BF_3, or $TiCl_4$ at a temperature range of −40 to 10 °C. These low molecular weight polyisobutylene polymers are used for caulking and as sealants. High molecular weight PIBs are synthesized at much lower temperatures of −100 to −90 °C and in a polar medium of methylene chloride. These polymers are used to make uncrosslinked rubber products and are additives for certain thermoplasts. Another characteristic of high molecular weight PIB is its low toxicity which allows it to be used as a base for chewing gum. The main chemical companies that produce polyisobutylene are Esso, ExxonMobil, and BASF.

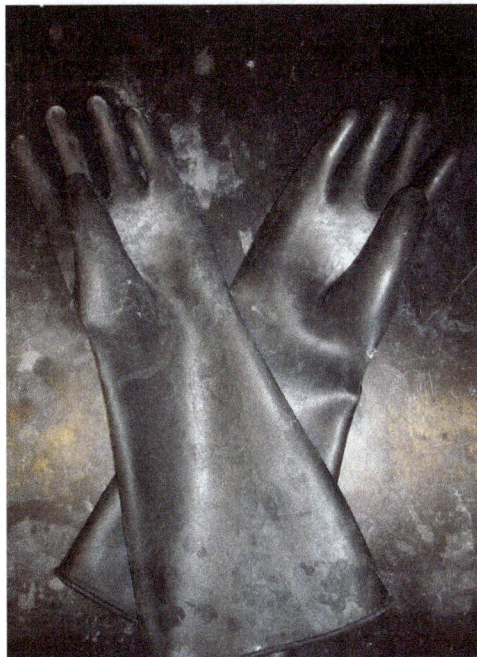
Butyl rubber gloves

Butyl rubber, in contrast to PIB, is a copolymer in which the monomers isobutylene (~98%) and isoprene (2%) are polymerized in a process similar to high molecular weight PIBs. Butyl rubber polymerization is carried out as a continuous process with $AlCl_3$ as the initiator. Its low gas permeability and good resistance to chemicals and aging make it useful for a variety of applications such as protective gloves, electrical cable insulation, and even basketballs. Large scale production of butyl rubber started during World War II, and roughly 1 billion pounds/year are produced in the U.S. today.

Polybutene is another copolymer, containing roughly 80% isobutylene and 20% other butenes (usually 1-butene). The production of these low molecular weight polymers (300–2500 Da) is done within a large range of temperatures (−45 to 80 °C) with $AlCl_3$ or BF_3. Depending on the molecular weight of these polymers, they can be used as adhesives, sealants, plasticizers, additives for transmission fluids, and a variety of other applications. These materials are low-cost and are made by a variety of different companies including BP Chemicals, Esso, and BASF.

Other polymers formed by cationic polymerization are homopolymers and copolymers of polyterpenes, such as pinenes (plant-derived products), that are used as tackyfiers. In the field of heterocycles, 1,3,5-trioxane is copolymerized with small amounts of ethylene oxide to form the highly crystalline polyoxymethylene plastic. Also, the homopolymerization of alkyl vinyl ethers is achieved only by cationic polymerization.

Living Polymerization

In polymer chemistry, living polymerization is a form of chain growth polymerization where the ability of a growing polymer chain to terminate has been removed. This can be accomplished in a variety of ways. Chain termination and chain transfer reactions are absent and the rate of chain initiation is also much larger than the rate of chain propagation. The result is that the polymer chains grow at a more constant rate than seen in traditional chain polymerization and their lengths remain very similar (i.e. they have a very low polydispersity index). Living polymerization is a popular method for synthesizing block copolymers since the polymer can be synthesized in stages, each stage containing a different monomer. Additional advantages are predetermined molar mass and control over end-groups.

Living polymerization is desirable because it offers precision and control in macromolecular synthesis. This is important since many of the novel/useful properties of polymers result from their microstructure and molecular weight. Since molecular weight and dispersity are less controlled in non-living polymerizations, this method is more desirable for materials design

In many cases, living polymerization reactions are confused or thought to be synonymous with controlled polymerizations. While these polymerization reactions are very similar, there is a distinct difference in the definitions of these two reactions. While living polymerizations are defined as polymerization reactions where termination or chain transfer is eliminated, controlled polymerization reactions are reactions where termination is suppressed, but not eliminated, through the introduction of a dormant state of the polymer. However, this distinction is still up for debate in the literature.

The main living polymerization techniques are:

- Living anionic polymerization

- Living cationic polymerization

- Living ring-opening metathesis polymerization

- Living free radical polymerization

- Living chain-growth polycondensations

History

Living polymerization was demonstrated by Michael Szwarc in 1956 in the anionic polymerization of styrene with an alkali metal / naphthalene system in tetrahydrofuran (THF) . After initial addition of monomer to the initiator system, the viscosity would increase (due to increased polymer chain growth), but eventually cease after depletion of monomer concentration. However, he found that addition of *more* monomer caused an increase in viscosity, indicating growth of the polymer chain, and thus concluded that the polymer chains had never been terminated. This was a major step in polymer chemistry, since control over when the polymer was quenched, or terminated, was generally not a controlled step. With this discovery, the list of potential applications expanded dramatically.

Today, living polymerizations are used widely in the development of various types of polymers or plastics. This is because chemists now have easy control of the chemical makeup of the polymer and, thus, the structural and electronic properties of the material. This level of control rarely exists in non-living polymerization reactions, leaving this method as the preferred synthetic route if attainable. The most common types of living polymerization reactions are anionic, cationic, free-radical, or ring-opening in nature, the specifics of each are discussed in the techniques section. In addition, while copolymers are possible to create using non-living polymerization reactions, the types of copolymers and the precise control of the chemical structure was expanded through the discovery of living polymerizations.

Characteristics

Fast Rate of Initiation

Another key characteristic is that the rate of initiation (meaning the dormant chemical species generates the active chain propagating species) must be much faster than the rate of chain propagation. This would allow all of the active species to form before chain propagation begins so all of the chains grow at the same rate (the rate of propagation). Conversely, if initiation is much slower than chain propagation (i.e. initiation is the rate determining step) then the active species will form at different points during the reaction leading to wider distribution between the individual polymer chains degree of polymerization (or chain length) and molecular weight.

Low Dispersity

Dispersity (Đ) or polydispersity index (PDI) is an indication of the broadness in the distribution of polymer chains. Living polymers tend to have low Đ due to the absence of chain termination pathways as well as the rate of initiation being much faster than the rate of propagation. If chain termination are present then chains will "die" or become inactive at various times during the polymerization which leads to polymer chains with varying x_n. As stated in the previous section, if the rate of propagation is much slower than the rate of initiation then all of the initiators will form the active species before the onset of propagation. When both of these characteristics are considered it becomes apparent that the concentration of active chains, that is those undergoing polymerization, becomes essentially constant. Both of these characteristics extend the lifetime of the propagating chain allowing for synthetic manipulation, co-block polymer formation and end group functionalization to be performed on the living chain. Increasing the lifetime of the propagating polymer chain allows for more control of the resulting polymers structure and properties since there is an inherent structure-property relationship in polymers.

Predictable Molecular Weight Since termination and chain transfer are absent in living polymerization then each initiator that generates an active species will be responsible for one chain. This offers control over the average degree of polymerization, which is related to M_n (number average molecular weight), in living polymerizations by controlling the monomer ($[M]_o$) to initiator ($[I]_o$) ratio. For an ideal living system, assuming efficiency for generating active species is 100%, where each initiator generates only one active species the Kinetic chain length (average number of monomers the active species reacts with during its lifetime) at a given time can be estimated by knowing the concentration of monomer remaining. The number average molecular weight, M_n, increases linearly with percent conversion during a living polymerization

$$v = \frac{[M]_0 - [M]}{[I]_0}$$

Living Polymerization Techniques

Living Anionic Polymerization

As early as 1936, Karl Ziegler proposed that anionic polymerization of styrene and butadiene by consecutive addition of monomer to an alkyl lithium initiator occurred without chain transfer or termination. Twenty years later, living polymerization was demonstrated by Szwarc through the anionic polymerization of styrene in THF using sodium naphthalenide as celerator.

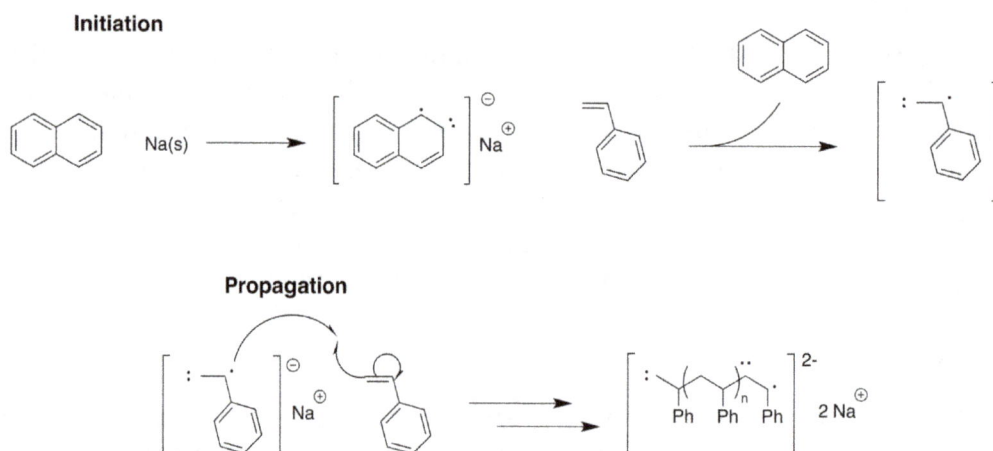

Initiation

Propagation

Here, the naphthalene anion acts as the initiator of the polymerization by activating the styrene. However, note that (with no impurities present for quenching and no solvent for chain transfer) there is no route for termination to occur. Therefore, these terminal anions will stay on the ends of the polymer until a quenching agent is introduced.

It is believed that the dianion of the polymer shown above is formed for this reaction, allowing the propagation to occur at either end of the chain. However, notice that there is no termination step (given impurities are not present to quench). This is the basis for anionic living polymerizations, where the terminal radical will exist until free monomer is available for additional propagation, or is quenched from an outside source.

Living α-olefin Polymerization

α-olefins can be polymerized through an anionic coordination polymerization in which the metal center of the catalyst is considered the counter cation for the anionic end of the alkyl chain (through a M-R coordination). Ziegler-Natta initiators were developed in the mid-1950s and are heterogeneous initiators used in the polymerization of alpha-olefins. Not only were these initiators the first to achieve relatively high molecular weight poly(1-alkenes) (currently the most widely produced thermoplastic in the world PE(Polyethylene) and PP (Polypropylene) but the initiators were also capable of stereoselctive polymerizations which is attributed to the chiral Crystal structure of the heterogeneous initiator. Due to the importance of this discovery Ziegler and Natta were presented with the 1963 Nobel Prize in chemistry. Although the active species formed from the Ziegler-Natta initiator generally have long lifetimes (on the scale of hours or longer) the lifetimes of the propagating chains are shortened due to several chain transfer pathways (Beta-Hydride elimination and transfer to the co-initiator) and as a result are not considered living.

Metallocene initiators are considered as a type of Ziegler-Natta initiators due to the use of the two-component system consisting of a Transition metal and a group I-III metal co-initiator (for example Methylalumoxane (MAO) or other alkyl aluminum compounds). The Metallocene initiators form homogenous single site catalysts that were initially developed to study the impact that the catalyst structure had on the resulting polymers structure/properties; which was difficult for multi-site heterogenous Ziegler-Natta initiators. Owing to the discrete single site on the metallocene catalyst researchers were able to tune and relate how the ancillary ligand (those not directly involved in the chemical transformations) structure and the symmetry about the chiral metal center affect the microstructure of the polymer. However, do to chain breaking reactions (mainly Beta-Hydride elimination) very few metallocene based polymerizations are known.

By tuning the steric bulk and electronic properties of the ancillary ligands and their substituents a class of initiators known as chelate initiators (or post-metallocene initiators) have been successfully used for stereospecific living polymerizations of alpha-olefins. The chelate initiators have a high potential for living polymerizations because the ancillary ligands can be designed to discourage or inhibit chain termination pathways. Chelate initiators can be further broken down based on the ancillary ligands; ansa-cyclopentyadienyl-amido initiators, alpha-diimine chelates and phenoxy-imine chelates.

- Ansa-cyclopentadienyl-amido (CpA) initiators

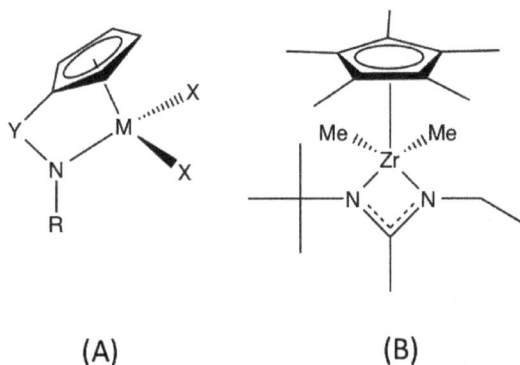

(A)   (B)

a.) Shows the general form of CpA initiators with one Cp ring and a coordinated Nitrogen b.) Shows the CpA initiator used in the living polymerization of 1-hexene (5)

CpA initiators have one Cyclopentadienyl substituent and one or more nitrogen substituents coordinated to the metal center (generally a Zr or Ti) (Odian). The dimethyl(pentamethylcyclopentyl) zirconium acetamidinate in figure____ has been used for a stereospecific living polymerization of 1-hexene at −10 deg C. The resulting poly(1-hexene) was isotactic (stereohemistry is the same between adjacent repeat units) confirmed by ^{13}C-NMR. The multiple trials demonstrated a controllable and predictable (from catalyst to monomer ratio) M_n with low Đ. The polymerization was further confirmed to be living by sequentially adding 2 portions of the monomer, the second portion was added after the first portion was already polymerized, and monitoring the Đ and M_n of the chain. The resulting polymer chains complied with the predicted M_n (with the total monomer concentration = portion 1 +2) and showed low Đ suggesting the chains were still active, or living, as the second portion of monomer was added (5).

- α-diimine chelate initiators

α-diimine chelate initiators are characterized by having a diimine chelating ancillary ligand structure and which is generally coordinated to a late transition (i.e. Ni and Pd) metal center.

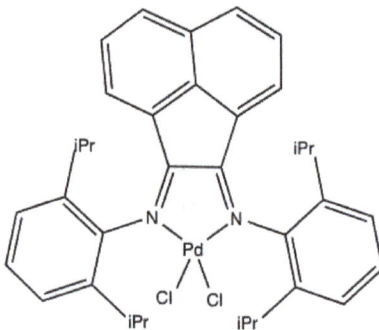

Brookhart et al. did extensive work with this class of catalysts and reported living polymerization for α-olefins and demonstrated living α-olefin carbon monoxide alternating copolymers.

- Phenoxy-imine chelates

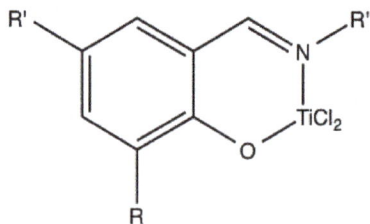

Living Cationic Polymerization

Monomers for living cationic polymerization are electron-rich alkenes such as vinyl ethers, isobutylene, styrene, and N-vinylcarbazole. The initiators are binary systems consisting of an electrophile and a Lewis acid. The method was developed around 1980 with contributions from Higashimura, Sawamoto and Kennedy. Typically, generating a stable carbocation for a prolonged period of time is difficult, due to the possibility for the cation to be quenched by a β-protons attached to another monomer in the backbone, or in a free monomer. Therefore a different approach is taken

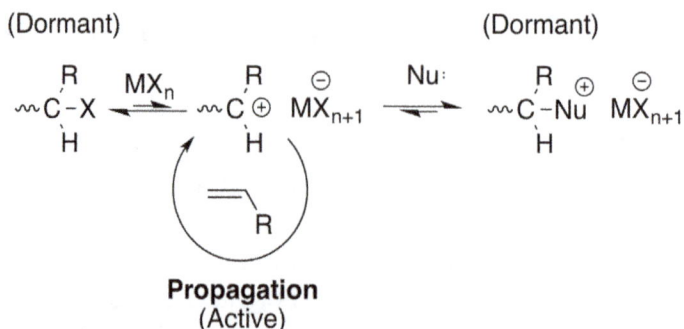

In this example, the carbocation is generated by the addition of a Lewis acid (co-initiator, along with the halogen "X" already on the polymer – see figure), which ultimately generates the carbocation in a weak equilibrium. This equilibrium heavily favors the dormant state, thus leaving little time for permanent quenching or termination by other pathways. In addition, a weak nucleophile (Nu:) can also added to reduce the concentration of active species even further, thus keeping the polymer "living". However, it is important to note that by definition, the polymers described in this example are not technically living due to the introduction of a dormant state, as termination has only been decreased, not eliminated (though this topic is still up for debate). But, they do operate similarly, and are used in similar applications to those of true living polymerizations.

Living Ring-opening Metathesis Polymerization

Given the right reaction conditions ring-opening metathesis polymerization (ROMP) can be rendered living. The first such systems were described by Robert H. Grubbs in 1986 based on norbornene and Tebbe's reagent and in 1978 Grubbs together with Richard R. Schrock describing living polymerization with a tungsten carbene complex.

Generally, ROMP reactions involve the conversion of a cyclic olefin with significant ring-strain (>5 kcal/mol), such as cyclobutene, norbornene, cyclopentene, etc., to a polymer that also contains double bonds. The important thing to note about ring-opening metathesis polymerizations is that the double bond is usually maintained in the backbone, which can allow it to be considered "living" under the right conditions.

For a ROMP reaction to be considered "living", several guidelines must be met:

1. Fast and complete initiation of the monomer. This means that the rate at which an initiating agent activates the monomer for polymerization, must happen very quickly.

2. How many monomers make up each polymer (the degree of polymerization) must be related linearly to the amount of monomer you started with.

3. The dispersity of the polymer must be < 1.5. In other words, the distribution of how long your polymer chains are in your reaction must be very low.

With these guidelines in mind, it allows you to create a polymer that is well controlled both in content (what monomer you use) and properties of the polymer (which can be largely attributed to polymer chain length). It is important to note that living ring-opening polymerizations can be anionic *or* cationic.

Because living polymers have had their termination ability removed, this means that once your monomer has been consumed, the addition of more monomer will result in the polymer chains continuing to grow until all of the additional monomer is consumed. This will continue until the metal catalyst at the end of the chain is intentionally removed by the addition of a quenching agent. As a result, it may potentially allow one to create a block or gradient copolymer fairly easily and accurately. This can lead to a high ability to tune the properties of the polymer to a desired application (electrical/ionic conduction, etc.)

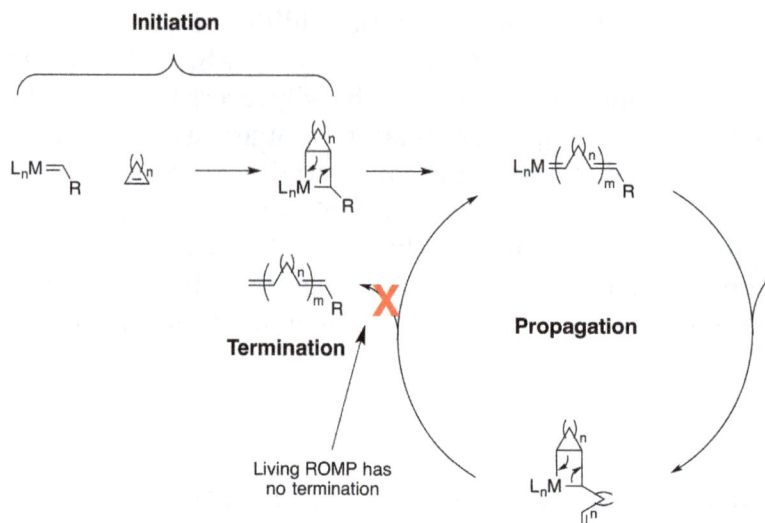

"Living" Free Radical Polymerization

Starting in the 1970s several new methods were discovered which allowed the development of living polymerization using free radical chemistry. These techniques involved catalytic chain transfer polymerization, iniferter mediated polymerization, stable free radical mediated polymerization (SFRP), atom transfer radical polymerization (ATRP), reversible addition-fragmentation chain transfer (RAFT) polymerization, and iodine-transfer polymerization.

In "living" radical polymerization (or controlled radical polymerization (CRP)) the chain breaking pathways are severely depressed when compared to conventional radical polymerization (RP) and CRP can display characteristics of a living polymerization. However, since chain termination is not absent, but only minimized, CRP technically does not meet the requirements imposed by IUPAC for a living polymerization. This issue has been up for debate the view points of different researchers can be found in a special issue of the Journal of Polymer Science titled Living or Controlled ?. The issue has not yet been resolved in the literature so it is often denoted as a "living" polymerization, quasi-living polymerization, pseudo-living and other terms to denote this issue.

There are two general strategies employed in CRP to suppress chain breaking reactions and promote fast initiation relative to propagation. Both strategies are based on developing a dynamic equilibrium amongst an active propagating radical and a dormant species.

The first strategy involves a reversible trapping mechanism in which the propagating radical undergoes an activation/deactivation (i.e. Atom-transfer radical-polymerization) process with a species X. The species X is a persistent radical, or a species that can generate a stable radical, that cannot terminate with itself or propagate but can only reversibly "terminate" with the propagating radical (from the propagating polymer chain)P*. P* is a radical species that an propagate (k_p) and irreversibly terminate (k_t) with another P*. X is normally a nitroxide (i.e. TEMPO used in Nitroxide Mediated Radical Polymerization) or an organometallic species. The dormant species (P_n-X) can be activated to regenerate the active propagating species (P*) spontaneously, thermally, using a catalyst and optically.

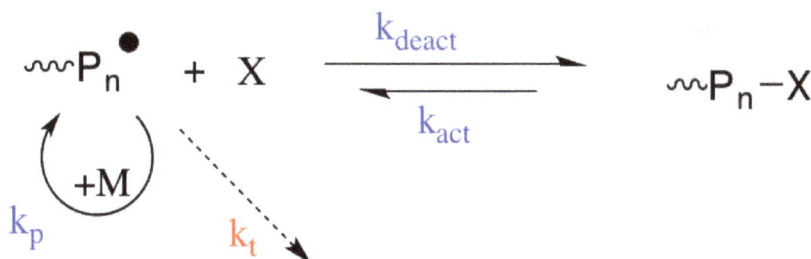

The second strategy is based on a degenerative transfer (DT) of the propagating radical between transfer agent that acts as a dormant species (i.e. Reversible addition–fragmentation chain-transfer polymerization). The DT based CRP's follow the conventional kinetics of radical polymerization, that is slow initiation and fast termination, but the transfer agent (Pm-X or Pn-X) is present in a much higher concentration compared to the radical initiator.The propagating radical species undergoes a thermally neutral exchange with the dormant transfer agent through atom transfer, group transfer or addition fragment chemistry.

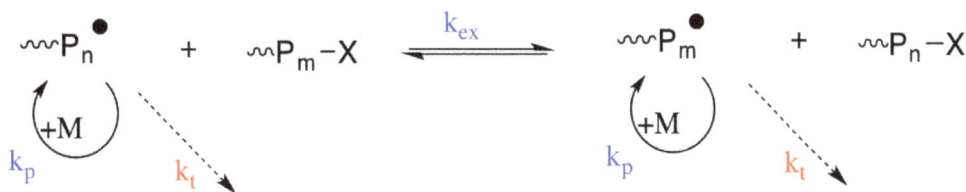

Living Chain-growth Polycondensations

Chain growth polycondensation polymerizations were initially developed under the premise that a change in substituent effects of the polymer, relative to the monomer, causes the polymers end group to be more reactive this has been referred to as "reactive intermediate polycondensation". The essential result is monomers preferentially react with the activated polymer end groups over reactions with other monomers. This preferred reactivity is the fundamental difference when categorizing a polymerization mechanism as chain-growth as opposed to step-growth in which the monomer and polymer chain end group have equal reactivity (the reactivity is uncontrolled). Several strategies were employed to minimize monomer-monomer reactions (or self-condensation) and polymerizations with low Đ and controllable Mn have been attained by this mechanism for small molecular weight polymers. However, for high molecular weight polymer chains (i.e. small initiator to monomer ratio) the Mn is not easily to controlled, for some monomers, since self-condensation between monomers occurred more frequently due to the low propagating species concentration.

Catalyst-transfer Polycondensation

Catalyst transfer polycondensation (CTP) is a chain-growth polycondensation mechanism in which the monomers do not directly react with one another and instead the monomer will only react with the polymer end group through a catalyst-mediated mechanism. The general process consists of the catalyst activating the polymer end group followed by a reaction of the end group with a 2nd incoming monomer. The catalyst is then transferred to the elongated chain while activating the end group.

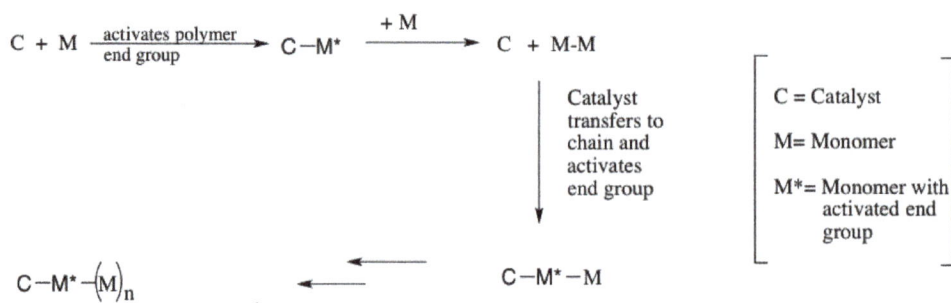

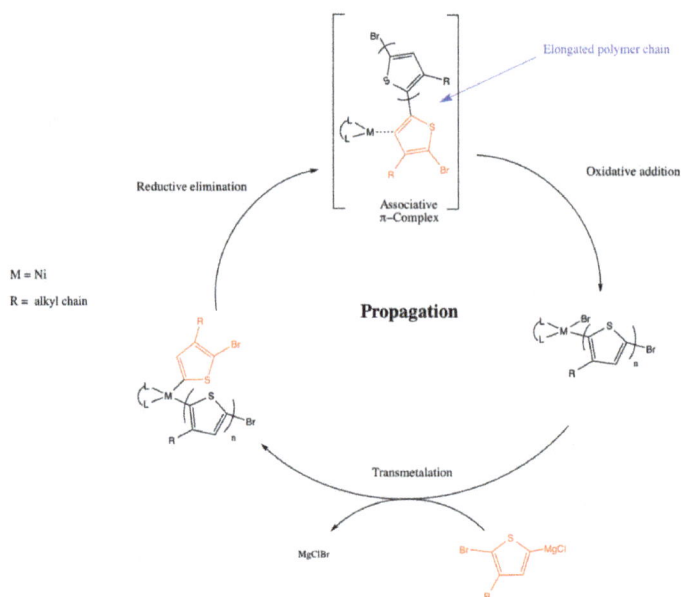

Catalyst transfer polycondensation allows for the living polymerization of π-conjugated polymers and was discovered by Tsutomu Yokozawa in 2004 and Richard McCullough. This was further refined by Ted Pappenfus in 2008 with the addition of a Grignard catalyst to initiate polymerization. In CTP the propagation step is based on organic cross coupling reactions (i.e. Kumada coupling, Sonogashira coupling, Negishi coupling) top form carbon carbon bonds between difunctional monomers. When Yokozawa and McCullough independently discovered the polymerization using a metal catalyst to couple a Grignard reagent with an organohalide making a new carbon-carbon bond. The mechanism below shows the formation of poly(3-alkylthiophene) using a Ni initiator (L_n can be 1,3-Bis(diphenylphosphino)propane (dppp)) and is similar to the conventional mechanism for Kumada coupling involving an oxidative addition, a transmetalation and a reductive elimination step. However there is a key difference, following reductive elimination in CTP, an associative complex is formed (which has been supported by intra-/intermolecular oxidative addition competition experiments) and the subsequent oxidative addition occurs between the metal center and the associated chain (an intramolecular pathway). Whereas in a coupling reaction the newly formed alkyl/aryl compound diffuses away and the subsequent oxidative addition occurs between an incoming Ar-Br bond and the metal center. The associative complex is essential to for polymerization to occur in a living fashion since it allows the metal to undergo a preferred intramolecular oxidative addition and remain with a single propagating chain (consistent with chain-growth mechanism), as opposed to an intermolecular oxidative addition with other monomers present in the solution (consistent with a step-growth, non-living, mechanism). The monomer scope of CTP has been increasing since its discovery and has included poly(phenylene)s, poly(fluorine)s, poly(selenophene)s and poly(pyrrole)s.

Living Group-transfer Polymerization

Group-transfer polymerization also has characteristics of living polymerization. It is applied to alkylated methacrylate monomers and the initiator is a silyl ketene acetal. New monomer adds to the initiator and to the active growing chain in a Michael reaction. With each addition of a monomer group the trimethylsilyl group is transferred to the end of the chain. The active chain-end is not ionic as in anionic or cationic polymeriation but is covalent. The reaction can be catalysed by bifluorides and bioxyanions such as *tris(dialkylamino) sulfonium bifluoride* or *tetrabutyl ammonium bibenzoate*. The method was discovered in 1983 by O.W. Webster and the name first suggested by Barry Trost.

Applications

Living polymerizations can be (and in some cases are) used industrially for many different applications. They can range from self-healing materials for space equipment to the easy design of copolymers for ion-exchange membranes in fuel cells, nanoscale lithography, etc.. While living polymerizations are still not widely used industrially, the field is rapidly growing, as well as the list of practical applications.

Self-healing Materials

Self-healing materials are materials in which repair, or "heal", themselves upon damage from an external force through the use of living polymers. For example, if a crack forms in the material, it proceeds to repair the crack and restore itself to its original, undamaged form. It achieves this by incorporating monomer-containing beads into a material made of a living polymer (with a terminally active chain). This has been achieved recently using a polyurethane derivative, with beads of monomer embedded in the material that become opened upon cracking of the material.

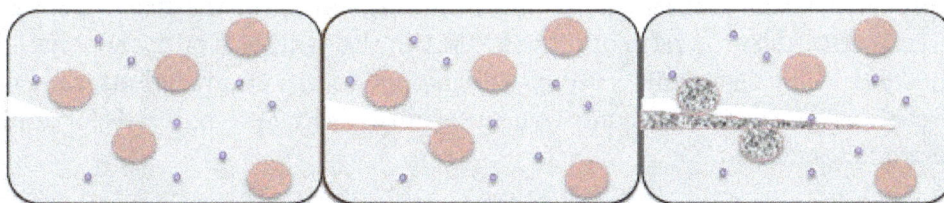

The polymer that makes up the material is designed as a living polymer, with reactive terminal end-groups that bind to the freshly provided monomer upon damage to the microbeads. This addition of monomer to the polymer chain increases the polymer chain to a length that fills the once open crack, in essence reconnecting all of the pieces back into one. According to Odriozola and coworkers, this application is originally designed for space equipment (in the event of debris damaging the equipment).

Copolymer Synthesis and Applications

Copolymers are polymers consisting of multiple different monomer species, and can be arranged in various orders, three of which are seen in the figure below.

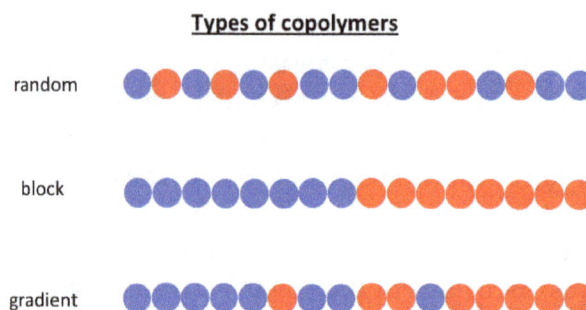

Types of copolymers

random

block

gradient

While there exist others (alternating copolymers, graft copolymers, and stereoblock copolymers), these three are more common in the scientific literature. In addition, block copolymers can exist as many types, including triblock (A-B-A), alternating block (A-B-A-B-A-B), etc.

Of these three types, block and gradient copolymers are commonly synthesized through living polymerizations, due to the ease of control living polymerization provides. Copolymers are highly desired due to the increased flexibility of properties a polymer can have compared to their homopolymer counterparts. The synthetic techniques used range from ROMP to generic anionic or cationic living polymerizations.

Copolymers, due to their unique tunability of properties, can have a wide range of applications. One example (of many) is nano-scale lithography using block copolymers. One used frequently is a block copolymer made of polystyrene and poly(methyl methacrylate) (abbreviated PS-*b*-PMMA). This copolymer, upon proper thermal and processing conditions, can form cylinders on the order of a few tens of nanometers in diameter of PMMA, surrounded by a PS matrix. These cylinders can then be etched away under high exposure to UV light and acetic acid, leaving a porous PS matrix.

The unique property of this material is that the size of the pores (or the size of the PMMA cylinders) can be easily tuned by the ratio of PS to PMMA in the synthesis of the copolymer. This can be easily tuned due to the easy control given by living polymerization reactions, thus making this technique highly desired for various nanoscale patterning of different materials for applications to catalysis, electronics, etc.

Reversible-deactivation Polymerization

Reversible-deactivation polymerization (RDP) is a form of polymerization propagated by chain carriers the majority of which at any instant are held in a state of dormancy through an equilibrium process involving other species. This ensures that the concentration of active carriers is sufficiently low as to render chain termination reactions negligible. Despite having some common features, it is distinct from living polymerization which requires a complete absence of termination.

Different Reversible-deactivation Polymerization

Atom-transfer radical-polymerization

Atom transfer radical polymerization (ATRP) is an example of a reversible-deactivation radical polymerization. Like its counterpart, ATRA or atom transfer radical addition, it is a means of forming a carbon-carbon bond through a transition metal catalyst. The polymerization from this method is called Atom transfer radical addition polymerization (ATRAP). As the name implies, the atom transfer step is the key step in the reaction responsible for uniform polymer chain growth. ATRP (or transition metal-mediated living radical polymerization) was independently discovered by Mitsuo Sawamoto and by Jin-Shan Wang and Krzysztof Matyjaszewski in 1995.

The following scheme presents a typical ATRP reaction:

General ATRP Reaction. A. Initiation. B. Equilibrium with dormant species. C.Propagation

Overview of ATRP

ATRP usually employs a transition metal complex as the catalyst with an alkyl halide as the initiator (R-X). Various transition metal complexes, namely those of Cu, Fe, Ru, Ni, Os, etc., have been employed as catalysts for ATRP. In an ATRP process, the dormant species is activated by the transition metal complex to generate radicals via one electron transfer process. Simultaneously the transition metal is oxidized to higher oxidation state. This reversible process rapidly establishes an equilibrium that is predominately shifted to the side with very low radical concentrations. The number of polymer chains is determined by the number of initiators. Each growing chain has the same probability to propagate with monomers to form living/dormant polymer chains (R-P$_n$-X). As a result, polymers with similar molecular weights and narrow molecular weight distribution can be prepared.

ATRP reactions are very robust in that they are tolerant of many functional groups like allyl, amino, epoxy, hydroxy and vinyl groups present in either the monomer or the initiator. ATRP methods are also advantageous due to the ease of preparation, commercially available and inexpensive catalysts (copper complexes), pyridine based ligands and initiators (alkyl halides).

The ATRP with styrene. If all the styrene is reacted (the conversion is 100%) the polymer will have 100 units of styrene built into it. PMDETA stands for N,N,N',N,N pentamethyldiethylenetriamine.

Components of Normal ATRP

There are four important variable components of Atom Transfer Radical Polymerizations. They are the monomer, initiator, catalyst and solvent. The following section breaks down the contributions of each component to the overall polymerization.

Monomer

Monomers that are typically used in ATRP are molecules with substituents that can stabilize the propagating radicals; for example, styrenes, (meth)acrylates, (meth)acrylamides, and acrylonitrile. ATRP are successful at leading to polymers of high number average molecular weight and a narrow polydispersity index when the concentration of the propagating radical balances the rate of radical termination. Yet, the propagating rate is unique to each individual monomer. Therefore, it is important that the other components of the polymerization (initiator, catalysts, ligands and solvents) are optimized in order for the concentration of the dormant species to be greater than the concentration of the propagating radical and yet not too great to slow down or halt the reaction.

Initiator

The number of growing polymer chains is determined by the initiator. Fast initiation ensures consistency of the number of propagating chains leading and to narrow molecular weight distributions. Organic halides that are similar in the organic framework as the propagating radical are often chosen as initiators. Most initiators for ATRP are alkyl halides. Alkyl halides such as alkyl bromides are more reactive than alkyl chlorides and both have good molecular weight control. The shape or structure of your initiator can determine the architecture of your polymer. For example, initiators with multiple alkyl halide groups on a single core can lead to a star-like polymer shape.

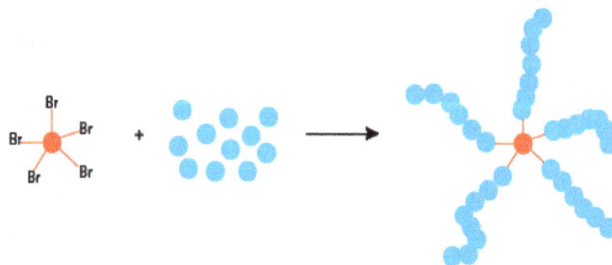

Illustration of a star initiator for ATRP

Catalyst

The catalyst is the most important component of ATRP because it determines the equilibrium constant between the active and dormant species. This equilibrium determines the polymerization rate and an equilibrium constant too small may inhibit or slow the polymerization while an equilibrium constant too large leads to a high distribution of chain lengths.

There are several requirements for the metal catalyst:

1. there needs to be two accessible oxidation states that are separated by one electron

2. the metal center needs to have a reasonable affinity for halogens

3. the coordination sphere of the metal needs to be expandable when it is oxidized so to be able to accommodate the halogen

4. the transition metal catalyst should not lead to significant side reactions, such as irreversible coupling with the propagating radicals and catalytic radical termination, etc.

The most studied catalysts are those that include copper, which has shown the most versatility, with successful polymerizations for a wide selection of monomers.

Solvents

Toluene,1,4-dioxane, xylene, anisole, DMF, DMSO, water, methanol, acetonitrile, and other solvents are used.

Kinetics of Normal ATRP

- Reactions in Atom Transfer Radical Polymerization

Initiation stage

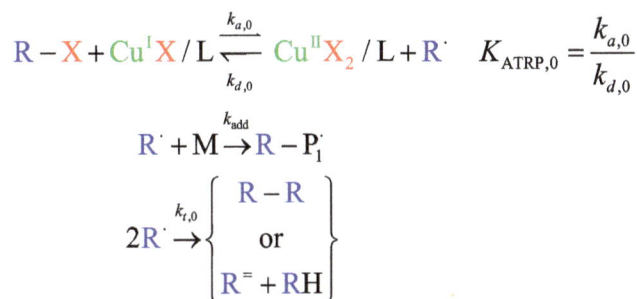

$$R-X+Cu^IX/L \underset{k_{d,0}}{\overset{k_{a,0}}{\rightleftharpoons}} Cu^{II}X_2/L+R^{\cdot} \qquad K_{ATRP,0}=\frac{k_{a,0}}{k_{d,0}}$$

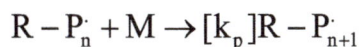

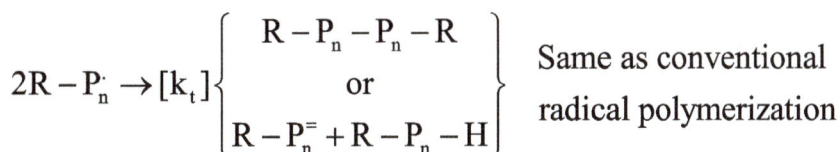

$$R^{\cdot}+M \xrightarrow{k_{add}} R-P_1^{\cdot}$$

$$2R^{\cdot} \xrightarrow{k_{t,0}} \begin{cases} R-R \\ \text{or} \\ R^{=}+RH \end{cases}$$

Quasi-steady state

ATRP
activation / deactivation

$$R-P_n-X+Cu^IX/L \underset{k_d}{\overset{k_a}{\rightleftharpoons}} Cu^{II}X2/L+R-P_n^{\cdot} \text{ equilibrium}$$

$$k_{ATRP}=\frac{k_a}{k_d}$$

$$R-P_n^{\cdot}+M \rightarrow [k_p]R-P_{n+1}^{\cdot}$$

$$2R-P_n^{\cdot} \rightarrow [k_t] \begin{cases} R-P_n-P_n-R \\ \text{or} \\ R-P_n^{=}+R-P_n-H \end{cases} \quad \begin{array}{l} \text{Same as conventional} \\ \text{radical polymerization} \end{array}$$

Other chain breaking reactions (k_{tx}) should also be considered.

ATRP Equilibrium Constant

The radical concentration in normal ATRP can be calculated via the following equation:

$$[R-P_n^{\bullet}] = K_{ATRP} \cdot [R-P_n-X] \cdot \frac{[Cu^I X / L]}{[Cu^{II} X2 / L]}$$

It is important to know the K_{ATRP} value to adjust the radical concentration. The K_{ATRP} value depends on the homo-cleavage energy of the alkyl halide and the redox potential of the Cu catalyst with different ligands. Given two alkyl halides (R¹-X and R²-X) and two ligands (L¹ and L²), there will be four combinations between different alkyl halides and ligands. Let K^{ij}_{ATRP} refer to the K_{ATRP} value for Rⁱ-X and Lʲ. If we know three of these four combinations, the fourth one can be calculated as $K^{22}_{ATRP} = K^{12}_{ATRP} \times K^{21}_{ATRP} / K^{11}_{ATRP}$. The K_{ATRP} values for different alkyl halides and different Cu catalysts can be found in literature.

Solvents have significant effects on the K_{ATRP} values. The K_{ATRP} value increases dramatically with the polarity of the solvent for the same alkyl halide and the same Cu catalyst. The polymerization must take place in solvent/monomer mixture, which changes to solvent/monomer/polymer mixture gradually. The K_{ATRP} values could change 10000 times by switching the reaction medium from pure methyl acrylate to pure dimethyl sulfoxide.

Activation and Deactivation Rate Coefficients

Deactivation rate coefficient, k_d, values must be sufficiently large to obtain low M_w/M_n value. The direct measurement of k_d value is difficult though not impossible. In most cases the k_d values were calculated from known K_{ATRP} and k_a. Cu complexes giving very low k_d values are not recommended to be used in ATRP reactions.

Retention of Chain End Functionality

$$\sum [X] = \text{Constant}$$

$$[R-X]_0 - [R-X]_t - [R-P_n-X]_t = ([Cu^I X / L]_t + 2[Cu^{II} X2 / L]_t) - ([Cu^I X / L]_0 + 2[Cu^{II} X2 / L]_0) + [RA-X]_t$$

Loss of chain end functionality / Change in $[Cu^I X/L]$ and $[Cu^{II} X2/L]$ / X transfer in activator regeneration

Halogen Conservation in Atom Transfer Radical Polymerization

High level retention of chain end functionality is desired. However, the determination of the loss of chain end functionality based on ¹H NMR and MS methods cannot provide precise values. As a result, it is difficult to identify the contributions of different chain breaking reactions in ATRP. There is a simple rule in ATRP which is the principle of halogen conservation. Halogen conservation means the total amount of halogen in the reaction systems must remain as a constant. Based on the simple rule, the level of retention of chain end functionality can be precisely determined in many cases. The precise determination of the loss of chain end functionality enabled further investigation of the chain breaking reactions in ATRP.

Different ATRP Methods

Activator Regeneration ATRP Methods

In a normal ATRP, the concentration of radicals is determined by the K_{ATRP} value, concentration of dormant species and $[Cu^I]/[Cu^{II}]$ ratio. In principle, the total amount of Cu catalyst should not influence the polymerization kinetics. However, the loss of chain end functionality slowly but irreversibly converts $[Cu^I]$ to $[Cu^{II}]$. Thus the initial $[Cu^I]$ is typically 0.1~1 equiv to the initiator. When very low concentrations of catalysts are used, usually at tens of ppm level, activator regeneration processes are generally required to compensate the loss of CEF and regenerate a sufficient amount of $[Cu^I]$ to continue the polymerization. Several activator regeneration ATRP methods were developed, namely ICAR ATRP, ARGET ATRP, SARA ATRP, eATRP and Photoinduced ATRP. The activator regeneration process is introduced to compensate the loss of chain end functionality, thus the cumulative amount of activator regeneration should roughly equal the total amount of the loss of chain end functionality.

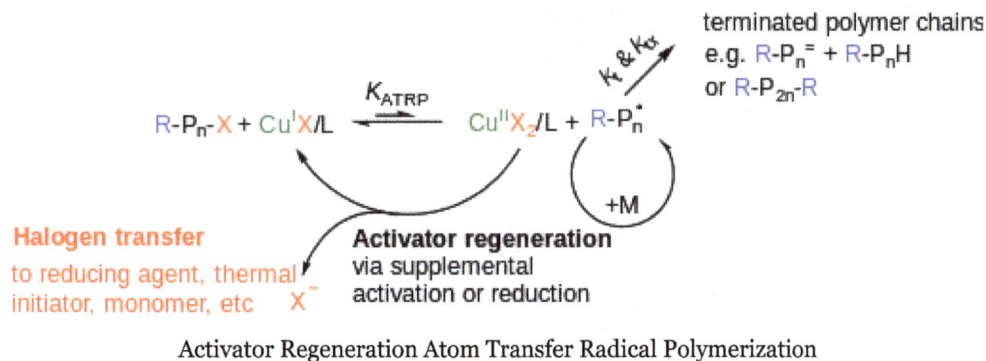

Activator Regeneration Atom Transfer Radical Polymerization

ICAR ATRP

Initiators for continuous activator regeneration (ICAR) is a technique that uses conventional radical initiators to continuously regenerate the activator, lowering its required concentration from thousands of ppm to <100 ppm; making it an industrially relevant technique.

ARGET ATRP

Activators regenerated by electron transfer (ARGET) employs non-radical forming reducing agents for regeneration of Cu^I. A good reducing agent (e.g. hydrazine, phenoles, sugars, ascorbic acid, etc...) should only react with Cu^{II} and not with radicals or other reagents in the reaction mixture.

SARA ATRP

A typical SARA ATRP employs Cu^0 as both supplemental activator and reducing agent (SARA). Cu^0 can activate alkyl halide directly but slowly. Cu^0 can also reduce Cu^{II} to Cu^I. Both processes help to regenerate Cu^I activator. Other zero valent metals, such as Mg, Zn and Fe, have also been employed for Cu-based SARA ATRP.

eATRP

In eATRP the activator Cu^I is regenerated via electrochemical process. The development of eATRP enables precise control of the reduction process and external regulation of the polymerization. In an eATRP process, the redox reaction involves two electrodes. The Cu^{II} species is reduced to Cu^I at the cathode. The anode compartment is typically separated from the polymerization environment, by using a glass frit and a conductive gel. Alternatively, a sacrificial aluminum counter electrode can be used, which is directly immersed in the reaction mixture.

Photoinduced ATRP

The direct photo reduction of transition metal catalysts in ATRP and/or photo assistant activation of alkyl halide is particularly interesting because such a procedure will allow performing of ATRP with ppm level of catalysts without any other additives.

Other ATRP Methods

Reverse ATRP

In reverse ATRP, the catalyst is added in its higher oxidation state. Chains are activated by conventional radical initiators (e.g. AIBN) and deactivated by the transition metal. The source of transferrable halogen is the copper salt, so this must be present in concentrations comparable to the transition metal.

SR&NI ATRP

A mixture of radical initiator and active (lower oxidation state) catalyst allows for the creation of block copolymers (contaminated with homopolymer) which is impossible using standard reverse ATRP. This is called SR&NI (simultaneous reverse and normal initiation ATRP).

AGET ATRP

Activators generated by electron transfer uses a reducing agent unable to initiate new chains (instead of organic radicals) as regenerator for the low-valent metal. Examples are metallic Cu, tin(II), ascorbic acid, or triethylamine. It allows for lower concentrations of transition metals, and may also be possible in aqueous or dispersed media.

Hybrid and Bimetallic Systems

This technique uses a variety of different metals/oxidation states, possibly on solid supports, to act as activators/deactivators, possibly with reduced toxicity or sensitivity. Iron salts can, for example, efficiently activate alkyl halides but requires an efficient Cu(II) deactivator which can be present in much lower concentrations (3–5 mol%)

Metal-free ATRP

Trace metal catalyst remaining in the final product has limited the application of ATRP in biomedical and electronic fields. In 2014, Craig Hawker and coworkers developed a new catalysis system involving photoredox reaction of 10-phenothiazine. The metal-free ATRP has been demonstrated to be capable of controlled polymerization of methacrylates. This technique was later expanded to polymerization of acrylonitrile by Krzysztof Matyjaszewski et al.

Polymers Made by ATRP

- Polystyrene

- Poly (methyl methacrylate)

- Polyacrylamide

Reversible-deactivation Radical Polymerization

Reversible deactivation radical polymerizations are members of the class of reversible deactivation polymerizations which exhibit much of the character of living polymerizations, but cannot be categorized as such as they are not without chain transfer or chain termination reactions. Several different names have been used in literature, which are:

- Living Radical Polymerization

- Living Free Radical Polymerization

- Controlled/"Living" Radical Polymerization

- Controlled Radical Polymerization

- Reversible Deactivation Radical Polymerization

Though the term "living" radical polymerization was used in early days, it has been discouraged by IUPAC, because radical polymerization cannot be a truly living process due to unavoidable termination reactions between two radicals. The commonly used term controlled radical polymerization is permitted, but reversible-deactivated radical polymerization or controlled reversible-deactivation radical polymerization (RDRP) is recommended.

History and Character

RDRP – sometimes misleadingly called 'free' radical polymerization – is one of the most widely used polymerization processes since it can be applied

- to a great variety of monomers

- it can be carried out in the presence of certain functional groups

- the technique is rather simple and easy to control

- the reaction conditions can vary from bulk over solution, emulsion, miniemulsion to suspension

- it is relatively inexpensive compared with competitive techniques

The steady-state concentration of the growing polymer chains is 10^{-7} M by order of magnitude, and the average life time of an individual polymer radical before termination is about 5-10 s. A drawback of the conventional radical polymerization is the limited control of chain architecture, molecular weight distribution, and composition. In the late 20th century it was observed that when certain components were added to systems polymerizing by a chain mechanism they are able to react reversibly with the (radical) chain carriers, putting them temporarily into a 'dormant' state. This had the effect of prolonging the lifetime of the growing polymer chains to values comparable with the duration of the experiment. At any instant most of the radicals are in the inactive (dormant) state, however, they are not irreversibly terminated ('dead'). Only a small fraction of them are active (growing), yet with a fast rate of interconversion of active and dormant forms, faster than the growth rate, the same probability of growth is ensured for all chains, i.e., on average, all chains are growing at the same rate. Consequently, rather than a most probable distribution, the molecular masses (degrees of polymerization) assume a much narrower Poisson distribution, and a lower dispersity prevails.

IUPAC also recognizes the alternative name, 'controlled reversible-deactivation radical polymerization' as acceptable, "provided the controlled context is specified, which in this instance comprises molecular mass and molecular mass distribution." These types of radical polymerizations are not necessarily 'living' polymerizations, since chain termination reactions are not precluded".

The adjective 'controlled' indicates that a certain kinetic feature of a polymerization or structural aspect of the polymer molecules formed is controlled (or both). The expression 'controlled polymerization' is sometimes used to describe a radical or ionic polymerization in which reversible-deactivation of the chain carriers is an essential component of the mechanism and interrupts the propagation that secures control of one or more kinetic features of the polymerization or one or more structural aspects of the macromolecules formed, or both. The expression 'controlled radical polymerization' is sometimes used to describe a radical polymerization that is conducted in the presence of agents that lead to e.g. atom-transfer radical polymerization (ATRP), nitroxide-(aminoxyl) mediated polymerization (NMP), or reversible-addition-fragmentation chain transfer (RAFT) polymerization. All these and further controlled polymerizations are included in the class of reversible-deactivation radical polymerizations. Whenever the adjective 'controlled' is used in this context the particular kinetic or the structural features that are controlled have to be specified.

Reversible-deactivation Polymerization

There is a mode of polymerization referred to as reversible-deactivation polymerization which is distinct from living polymerization, despite some common features. Living polymerization requires a complete absence of termination reactions, whereas reversible-deactivation polymerization may contain a similar fraction of termination as conventional polymerization with the same concentration of active species. Some important aspects of these are compared in the table:

Comparison of radical polymerization processes			
Property	Standard radical polymerization	Living polymerization	Reversible-deactivation polymerization
Concn. of initiating species	Falls off only slowly	Falls off very rapidly	Falls off very rapidly
Concn. of chain carriers (Number of growing chains)	Instantaneous steady state (Bodenstein approximation applies) decreasing throughout reaction	Constant throughout reaction	Constant throughout reaction
Lifetime of growing chains	$\sim 10^{-3}$ s	Same as reaction duration	Same as reaction duration
Main form of termination	Radical combination or radical disproportionation	Termination reactions are precluded	Termination reactions are not precluded
Degree of polymerization	Broad range (Đ >=1.5), Schulz-Zimm distribution	Narrow range(Đ <1.5), Poisson distribution	Narrow range(Đ <1.5), Poisson distribution
Dormant states	None	Rare	Predominant

Common Features of RDRP

As the name suggests, the prerequisite of a successful RDRP is fast and reversible activation/deactivation of propagating chains. There are three types of RDRP; namely deactivation by catalyzed reversible coupling, deactivation by spontaneous reversible coupling and deactivation by degenerative transfer (DT). A mixture of different mechanisms is possible; e.g. a transition metal mediated RDRP could switch among ATRP, OMRP and DT mechanisms depending on the reaction conditions and reagents used.

In any RDRP processes, the radicals can propagate with the rate coefficient k_p by addition of a few monomer units before the deactivation reaction occurs to regenerate the dormant species. Concurrently, two radicals may react with each other to form dead chains with the rate coefficient k_t. The rates of propagation and termination between two radicals are not influenced by the mechanism of deactivation or the catalyst used in the system. Thus it is possible to estimate how fast a RDRP can be conducted with preserved chain end functionality?

In addition, other chain breaking reactions such as irreversible chain transfer/termination reactions of the propagating radicals with solvent, monomer, polymer, catalyst, additives, etc. would introduce additional loss of chain end functionality (CEF). The overall rate coefficient of chain breaking reactions besides the direct termination between two radicals is represented as k_{tx}.

1) Deactivation by catalyzed reversible coupling
e.g. ATRP

$$R\text{-}P_n\text{-}X + Mt^n/L \underset{k_d}{\overset{k_a}{\rightleftharpoons}} R\text{-}P_n^\bullet + X\text{-}Mt^{n+1}/L \qquad \boxed{X = Cl,\ Br,\ etc.}$$

activator   deactivator

2) Deactivation by spontaneous reversible coupling
e.g. NMP, OMRP

$$R\text{-}P_n\text{-}X \underset{k_d}{\overset{k_a}{\rightleftharpoons}} R\text{-}P_n^\bullet + X \qquad \boxed{X = \ ^\bullet O\text{-}N \quad , etc.}$$

3) Dactivation by degenerative transfer
e.g. RAFT, Unsaturated polymethacrylates,
I, Te, Ge, Sn, Sb, Bi, etc. mediated radical polymerization

$$R\text{-}P_n\text{-}X + R\text{-}P_m^\bullet \underset{k_{tr}}{\overset{k_{tr}}{\rightleftharpoons}} R\text{-}P_n^\bullet + R\text{-}P_m\text{-}X \qquad \boxed{-X = -S-\overset{S}{\overset{\|}{C}}-Z \quad , etc.}$$

Propagation and chain breaking reactions

$$R\text{-}P_n^\bullet \xrightarrow{k_t\ \&\ k_{tx}} \text{terminated polymer chains}$$

$$\circlearrowleft\ +M\ ,\ k_p$$

e.g. $R\text{-}P_n^= + R\text{-}P_n H$

or $R\text{-}P_n\text{-}P_n\text{-}R$

In all RDRP methods, the theoretical number average molecular weight of obtained polymers, M_n, can be defined by following equation:

$$M_n = M_m \times \frac{[M]_0 - [M]_t}{[R\text{-}X]_0}$$

where M_m is the molecular weight of monomer; $[M]_0$ and $[M]_t$ are the monomer concentrations at time 0 and time t; $[R\text{-}X]_0$ is the initial concentration of the initiator.

Besides the designed molecular weight, a well controlled RDRP should give polymers with narrow molecular distributions, which can be quantified by M_w/M_n values, and well preserved chain end functionalities.

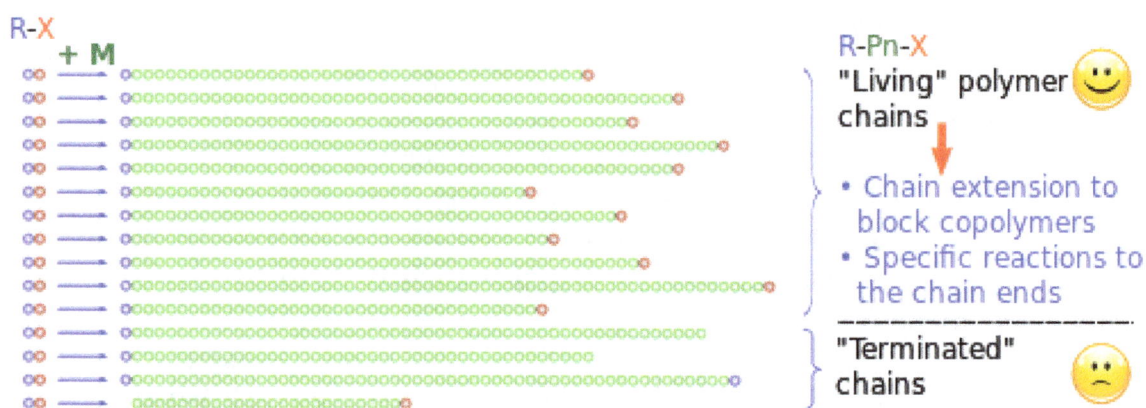

A well controlled RDRP process requires: 1) the reversible deactivation process should be sufficiently fast; 2) the chain breaking reactions which cause the loss of chain end functionalities should be limited; 3) properly maintained radical concentration; 4) the initiator should have proper activity.

Examples of RDRP

Atom-transfer Radical Polymerization (ATRP)

The initiator of the polymerization is usually an organohalogenid and the dormant state is achieved in a metal complex of a transition metal ('radical buffer'). This method is very versatile but requires unconventional initiator systems that are sometimes poorly compatible with the polymerization media.

Nitroxide-mediated Polymerization (NMP)

Given certain conditions a homolytic splitting of the C-O bond in alkoxylamines can occur and a stable 2-centre 3 electron N-O radical can be formed that is able to initiate a polymerization reaction. The preconditions for an alkoxylamine suitable to initiate a polymerization are bulky, sterically obstructive substituents on the secondary amine, and the substituent on the oxygen should be able to form a stable radical, e.g. benzyl.

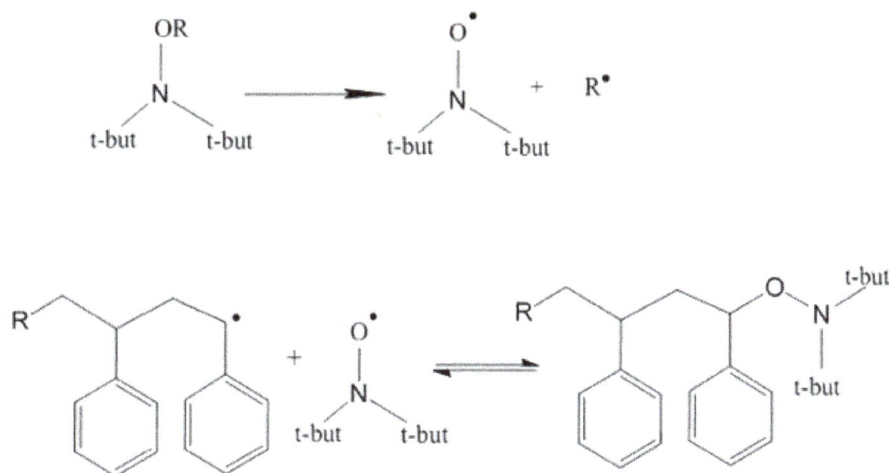

Example of a reversible deactivation reaction

Reversible Addition-fragmentation Chain Transfer (RAFT)

RAFT is one of the most versatile and convenient techniques in this context. The most common RAFT-processes are carried out in the presence of thiocarbonylthio compounds that act as radical buffers. While in ATRP and NMP reversible deactivation of propagating radical-radical reactions takes place and the dormant structures are a halo-compound in ATRP and the alkoxyamine in NMP, both being a sink for radicals and source at the same time and described by the corresponding equilibria. RAFT on the contrary, is controlled by chain-transfer reactions that are in a deactivation-activation equilibrium. Since no radicals are generated or destroyed an external source of radicals is necessary for initiation and maintenance of the propagation reaction.

Initiation step of a RAFT polymerization

$$I \rightarrow I^{\cdot} \xrightarrow{M} \xrightarrow{M} P_n^{\cdot}$$

Reversible chain transfer

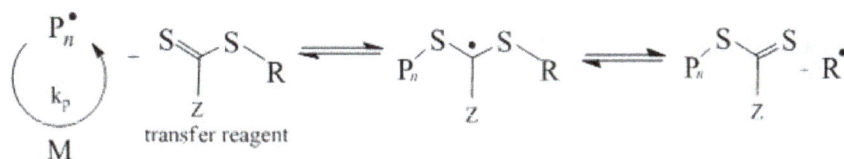

Reinitiation step

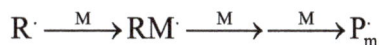

$$R^{\cdot} \xrightarrow{M} RM^{\cdot} \xrightarrow{M} \xrightarrow{M} P_m^{\cdot}$$

Chain equilibration step

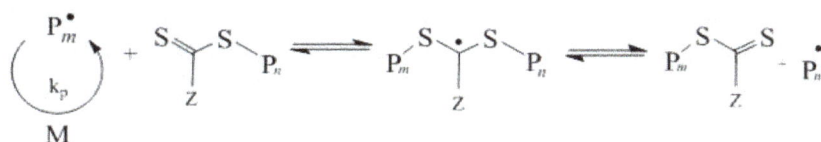

Termination step

$$P_m^{\cdot} + P_n^{\cdot} \rightarrow P_m P_n$$

Catalytic Chain Transfer and Cobalt Mediated Radical Polymerization

Although not a strictly living form of polymerization catalytic chain transfer polymerization must be mentioned as it figures significantly in the development of later forms of living free radical polymerization. Discovered in the late 1970s in the USSR it was found that cobalt porphyrins were able to reduce the molecular weight during polymerization of methacrylates. Later investigations showed that the cobalt glyoxime complexes were as effective as the porphyrin catalysts and also less oxygen sensitive. Due to their lower oxygen sensitivity these catalysts have been investigated much more thoroughly than the porphyrin catalysts.

The major products of catalytic chain transfer polymerization are vinyl terminated polymer chains. One of the major drawbacks of the process is that catalytic chain transfer polymerization does not produce macromonomers but instead produces addition fragmentation agents. When a growing polymer chain reacts with the addition fragmentation agent the radical end-group attacks the vinyl bond and forms a bond. However, the resulting product is so hindered that the species undergoes fragmentation, leading eventually to telechelic species.

These addition fragmentation chain transfer agents do form graft copolymers with styrenic and acrylate species however they do so by first forming block copolymers and then incorporating these block copolymers into the main polymer backbone.

While high yields of macromonomers are possible with methacrylate monomers, low yields are obtained when using catalytic chain transfer agents during the polymerization of acrylate and stryenic monomers. This has been seen to be due to the interaction of the radical centre with the catalyst during these polymerization reactions.

The reversible reaction of the cobalt macrocycle with the growing radical is known as cobalt carbon bonding and in some cases leads to living polymerization reactions.

Iniferter Polymerization

An iniferter is a chemical compound that simultaneously acts as initiator, transfer agent, and terminator (hence the name ini-fer-ter) in controlled free radical iniferter polymerizations, the most common is the dithiocarbamate type.

Iodine-Transfer Polymerization (ITP)

Iodine-transfer polymerization (ITP, also called ITRP), developed by Tatemoto and coworkers in the 1970s gives relatively low polydispersities for fluoroolefin polymers. While it has received relatively little academic attention, this chemistry has served as the basis for several industrial patents and products and may be the most commercially successful form of living free radical polymerization. It has primarily been used to incorporate iodine cure sites into fluoroelastomers.

$$In_2 \longrightarrow In\bullet \longrightarrow P_1\bullet \xrightarrow{\;M\;} P_m\bullet \underset{k_p}{\overset{M}{\circlearrowright}}$$

$$P_m\bullet \;+\; I\text{-}R \rightleftharpoons P_m\text{-}I \;+\; R^{\bullet}$$

$$R\bullet \xrightarrow{\;M\;} P_n\bullet \underset{k_p}{\overset{M}{\circlearrowright}}$$

$$\underset{k_p}{\overset{M}{\circlearrowright}} P_n\bullet \;+\; I\text{-}P_m \rightleftharpoons P_n\text{-}I \;+\; P_m\bullet \underset{k_p}{\overset{M}{\circlearrowright}}$$

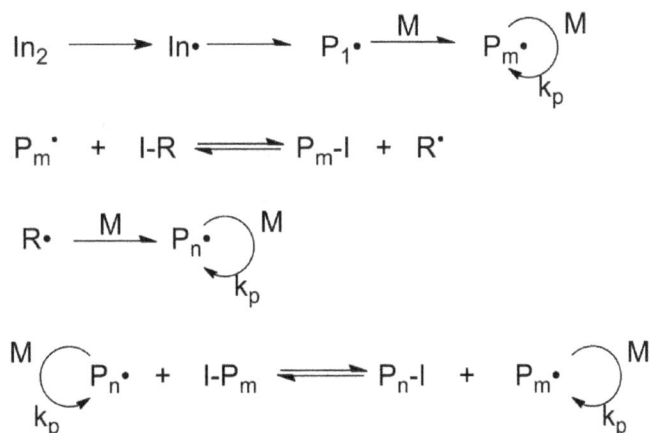

The mechanism of ITP involves thermal decomposition of the radical initiator (typically persulfate), generating the initiating radical In•. This radical adds to the monomer M to form the species $P_1\bullet$, which can propagate to $P_m\bullet$. By exchange of iodine from the transfer agent R-I to the propagating radical $P_m\bullet$ a new radical R• is formed and $P_m\bullet$ becomes dormant. This species can propagate with monomer M to $P_n\bullet$. During the polymerization exchange between the different polymer chains and the transfer agent occurs, which is typical for a degenerative transfer process.

Typically, iodine transfer polymerization uses a mono- or diiodo-perfluoroalkane as the initial chain transfer agent. This fluoroalkane may be partially substituted with hydrogen or chlorine. The energy of the iodine-perfluoroalkane bond is low and, in contrast to iodo-hydrocarbon bonds, its polarization small. Therefore, the iodine is easily abstracted in the presence of free radicals. Upon encountering an iodoperfluoroalkane, a growing poly(fluoroolefin) chain will abstract the iodine and terminate, leaving the now-created perfluoroalkyl radical to add further monomer. But the iodine-terminated poly(fluoroolefin) itself acts as a chain transfer agent. As in RAFT processes, as long as the rate of initiation is kept low, the net result is the formation of a monodisperse molecular weight distribution.

Use of conventional hydrocarbon monomers with iodoperfluoroalkane chain transfer agents has been described. The resulting molecular weight distributions have not been narrow since the energetics of an iodine-hydrocarbon bond are considerably different from that of an iodine-fluorocarbon bond and abstraction of the iodine from the terminated polymer difficult. The use of hydrocarbon iodides has also been described, but again the resulting molecular weight distributions were not narrow.

Preparation of block copolymers by iodine-transfer polymerization was also described by Tatemoto and coworkers in the 1970s.

Although use of living free radical processes in emulsion polymerization has been characterized as difficult, all examples of iodine-transfer polymerization have involved emulsion polymerization. Extremely high molecular weights have been claimed.

Listed below are some other less described but to some extent increasingly important living radical polymerization techniques.

Selenium-Centered Radical-Mediated Polymerization

Diphenyl diselenide and several benzylic selenides have been explored by Kwon *et al.* as photoiniferters in polymerization of styrene and methyl methacrylate. Their mechanism of control over polymerization is proposed to be similar to the dithiuram disulfide iniferters. However, their low transfer constants allow them to be used for block copolymer synthesis but give limited control over the molecular weight distribution.

Telluride-Mediated Polymerization (TERP)

Telluride-Mediated Polymerization or TERP first appeared to mainly operate under a reversible chain transfer mechanism by homolytic substitution under thermal initiation. However, in a kinetic study it was found that TERP predominantly proceeds by degenerative transfer rather than 'dissociation combination'.

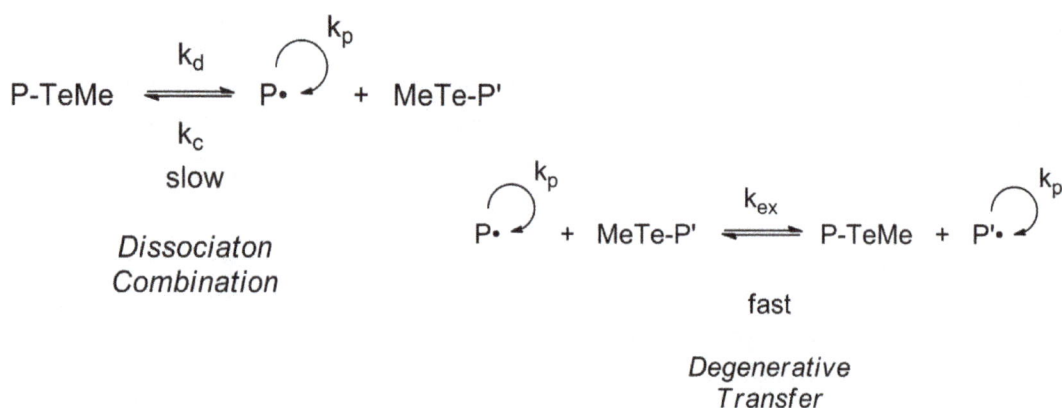

Alkyl tellurides of the structure Z-X-R, were Z=methyl and R= a good free radical leaving group, give the better control for a wide range of monomers, phenyl tellurides (Z=phenyl) giving poor control. Polymerization of methyl methacrylates are only controlled by ditellurides. The importance of X to chain transfer increases in the series O<S<Se<Te, makes alkyl tellurides effective in mediating control under thermally initiated conditions and the alkyl selenides and sulfides effective only under photoinitiated polymerization.

Stibine-Mediated Polymerization

More recently Yamago *et al.* reported stibine-mediated polymerization, using an organostibine transfer agent with the general structure Z(Z')-Sb-R (where Z= activating group and R= free radical leaving group). A wide range of monomers (styrenics, (meth)acrylics and vinylics) can be controlled, giving narrow molecular weight distributions and predictable molecular weights under thermally initiated conditions. Yamago has also published a patent indicating that bismuth alkyls can also control radical polymerizations via a similar mechanism.

References

- Susan E. M. Selke, John D. Culter, Ruben J. Hernandez, "Plastics packaging: Properties, processing, applications, and regulations", Hanser, 2004, p.29. ISBN 1-56990-372-7

- Susan E. M. Selke, John D. Culter, Ruben J. Hernandez, "Plastics packaging: Properties, processing, applications, and regulations", Hanser, 2004, p.29. ISBN 1-56990-372-7

- Odian, George (2004). Principles of Polymerization (4th ed.). Hoboken, NJ: Wiley-Interscience. ISBN 978-0-471-27400-1.

- Mark, Herman F.; Bikales, Norbert; Overberger, Charles G.; Menges, Georg; Kroschwitz, Jacqueline I. (1990). Encyclopedia of Polymer Science and Engineering (2nd ed.). Wiley-Interscience. ISBN 978-0-471-80950-0.

- Matyjaszewski, Krzysztof (1996). Cationic Polymerizations: Mechanisms, Synthesis, and Applications. New York, New York: Marcel Dekker, Inc. ISBN 978-0-8247-9463-7.

- Cowie, John M. G.; Arrighi, Valeria (2008). Polymers Chemistry and Physics of Modern Materials (3rd ed.). Boca Raton: Taylor & Francis. ISBN 978-0-8493-9813-1.

- Raave, A. (2000). Principles of Polymer Chemistry (2nd ed.). New York, New York: Kluwer Academic/Plenum Publishers. ISBN 978-0-306-46368-6.

- Ebewele, Robert Oboigbaotor (2000). Polymer Science and Technology. Boca Ration, FL: Chapman & Hall/CRC Press LLC. ISBN 978-0-8493-8939-9.

- Chanda, Manas; Roy, Salil K. (2007). Plastics Technology Handbook: Plastics Engineering Series (4th ed.). Boca Raton, FL: CSC Press. ISBN 978-0-8493-7039-7.

- Cowie, J.M.G. (2007). Polymers chemistry and physics of modern materials (3rd ed / J.M.G. Cowie and Valeria Arrighi ed.). Boca Raton: Taylor & Francis. ISBN 9780849398131.

- Odian, George (2004). Principles of polymerization (4. ed.). Hoboken, NJ: Wiley-Interscience. ISBN 0471274003.

- Cowie, J. M. G.; Arrighi, V. In Polymers: Chemistry and Physics of Modern Materials; CRC Press Taylor and Francis Group: Boca Raton, Fl, 2008; 3rd Ed., pp. 82–84 ISBN 0849398134

Degradation of Polymer

Polymer degradation is the change in the properties of a polymer. Some of these properties are the shape, tensile strength and shape of a polymer. These variations are usually unwanted. Some other aspects of polymer degradation listed in this chapter are UV degradation, thermal degradation of polymers, thermal depolymerization and photo-oxidation of polymers.

Polymer Degradation

Polymer degradation is a change in the properties—tensile strength, color, shape, etc.—of a polymer or polymer-based product under the influence of one or more environmental factors such as heat, light or chemicals such as acids, alkalis and some salts. These changes are usually undesirable, such as cracking and chemical disintegration of products or, more rarely, desirable, as in biodegradation, or deliberately lowering the molecular weight of a polymer for recycling. The changes in properties are often termed "aging".

In a finished product such a change is to be prevented or delayed. Degradation can be useful for recycling/reusing the polymer waste to prevent or reduce environmental pollution. Degradation can also be induced deliberately to assist structure determination.

Polymeric molecules are very large (on the molecular scale), and their unique and useful properties are mainly a result of their size. Any loss in chain length lowers tensile strength and is a primary cause of premature cracking.

Commodity Polymers

Today there are primarily seven commodity polymers in use: polyethylene, polypropylene, polyvinyl chloride, polyethylene terephthalate, polystyrene, polycarbonate, and poly(methyl methacrylate) (Plexiglas). These make up nearly 98% of all polymers and plastics encountered in daily life. Each of these polymers has its own characteristic modes of degradation and resistances to heat, light and chemicals. Polyethylene, polypropylene, and poly(methyl methacrylate) are sensitive to oxidation and UV radiation, while PVC may discolor at high temperatures due to loss of hydrogen chloride gas, and become very brittle. PET is sensitive to hydrolysis and attack by strong acids, while polycarbonate depolymerizes rapidly when exposed to strong alkalis.

For example, polyethylene usually degrades by *random scission*—that is by a random breakage of the linkages (bonds) that hold the atoms of the polymer together. When this polymer is heated above 450 Celsius it becomes a complex mixture of molecules of various sizes that resemble gasoline. Other polymers—like polyalphamethylstyrene—undergo 'specific' chain scission with breakage occurring only at the ends; they literally unzip or depolymerize to become the constituent monomers.

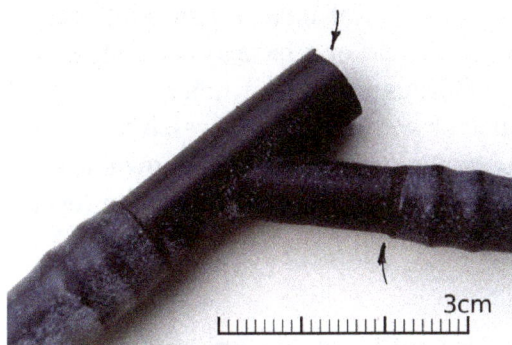

Close-up of broken fuel pipe from road traffic accident

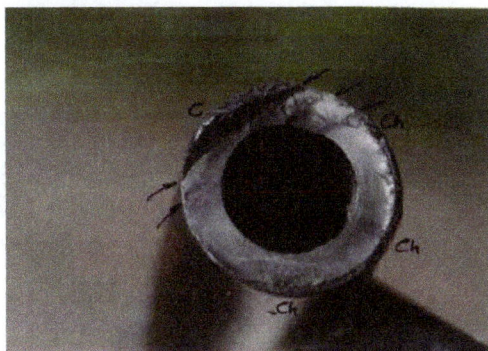

Close-up of broken fuel pipe connector

Photoinduced Degradation

Most polymers can be degraded by photolysis to give lower molecular weight molecules. Electromagnetic waves with the energy of visible light or higher, such as ultraviolet light, X-rays and gamma rays are usually involved in such reactions.

Thermal Degradation

Chain-growth polymers like poly(methyl methacrylate) can be degraded by thermolysis at high temperatures to give monomers, oils, gases and water. The degradation takes place by:

Thermolysis type	Added material	Temperature	Pressure	Final product
Pyrolysis		Around 500 °C	Reduced pressure	
Hydrogenation	Dihydrogen	Around 450 °C	Around 200 bars	
Gasification	Dioxygen and/or water		Under pressure	Carbon monoxide, Carbon dioxide and hydrogen

Chemical Degradation

Solvolysis

Step-growth polymers like polyesters, polyamides and polycarbonates can be degraded by solvolysis and mainly hydrolysis to give lower molecular weight molecules. The hydrolysis takes place in the presence of water containing an acid or a base as catalyst. Polyamide is sensitive to degradation by acids and polyamide mouldings will crack when attacked by strong acids. For example, the fracture surface of a fuel connector showed the progressive growth of the crack from acid attack (Ch) to the final cusp (C) of polymer. The problem is known as stress corrosion cracking, and in this case was caused by hydrolysis of the polymer. It was the reverse reaction of the synthesis of the polymer:

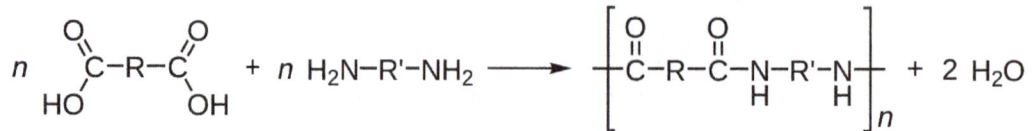

$$n \; \underset{HO}{\overset{O}{\underset{\|}{C}}}-R-\underset{OH}{\overset{O}{\underset{\|}{C}}} \; + \; n \; H_2N-R'-NH_2 \; \longrightarrow \; \left[\underset{}{\overset{O}{\underset{\|}{C}}}-R-\underset{}{\overset{O}{\underset{\|}{C}}}-\underset{H}{N}-R'-\underset{H}{N} \right]_n \; + \; 2 \; H_2O$$

Ozonolysis

Ozone cracking in Natural rubber tubing

Cracks can be formed in many different elastomers by ozone attack. Tiny traces of the gas in the air will attack double bonds in rubber chains, with Natural rubber, polybutadiene, Styrene-butadiene rubber and NBR being most sensitive to degradation. Ozone cracks form in products under tension, but the critical strain is very small. The cracks are always oriented at right angles to the strain axis, so will form around the circumference in a rubber tube bent over. Such cracks are dangerous when they occur in fuel pipes because the cracks will grow from the outside exposed surfaces into the bore of the pipe, and fuel leakage and fire may follow. The problem of ozone cracking can be prevented by adding anti-ozonants to the rubber before vulcanization. Ozone cracks were commonly seen in automobile tire sidewalls, but are now seen rarely thanks to these additives. On the other hand, the problem does recur in unprotected products such as rubber tubing and seals.

Oxidation

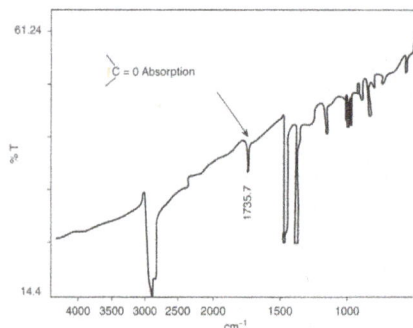

IR spectrum showing carbonyl absorption due to oxidative degradation of polypropylene crutch moulding

The polymers are susceptible to attack by atmospheric oxygen, especially at elevated temperatures encountered during processing to shape. Many process methods such as extrusion and injection moulding involve pumping molten polymer into tools, and the high temperatures needed for melting may result in oxidation unless precautions are taken. For example, a forearm crutch suddenly snapped and the user was severely injured in the resulting fall. The crutch had fractured across a polypropylene insert within the aluminium tube of the device, and infra-red spectroscopy of the material showed that it had oxidised, possible as a result of poor moulding.

Oxidation is usually relatively easy to detect owing to the strong absorption by the carbonyl group in the spectrum of polyolefins. Polypropylene has a relatively simple spectrum with few peaks at the carbonyl position (like polyethylene). Oxidation tends to start at tertiary carbon atoms because the free radicals formed here are more stable and longer lasting, making them more susceptible to attack by oxygen. The carbonyl group can be further oxidised to break the chain, this weakens the material by lowering its molecular weight, and cracks start to grow in the regions affected.

Galvanic Action

Polymer degradation by galvanic action was first described in the technical literature in 1990. This was the discovery that "plastics can corrode", i.e. polymer degradation may occur through galvanic action similar to that of metals under certain conditions and has been referred to as the "Faudree Effect". In the aerospace field, this finding has largely contributed to aircraft safety, mainly those aircraft that use CFRP and has resulted in a wide body of follow-up research and patents. Normally, when two dissimilar metals such as copper (Cu) and iron (Fe) are put into contact and then immersed in salt water, the iron will undergo corrosion, or rust. This is called a galvanic circuit where the copper is the noble metal and the iron is the active metal, i.e., the copper is the positive (+) electrode and the iron is the negative (-) electrode. A battery is formed. It follows that plastics are made stronger by impregnating them with thin carbon fibers only a few micrometers in diameter known as carbon fiber reinforced polymers (CFRP). This is to produce materials that are high strength and resistant to high temperatures. The carbon fibers act as a noble metal similar to gold (Au) or platinum (Pt). When put into contact with a more active metal, for example with aluminum (Al) in salt water the aluminum corrodes. However, in early 1990, it was reported that imide-linked resins in CFRP composites degrade when bare composite is coupled with an active metal in salt water environments. This is because corrosion not only occurs at the aluminum anode, but also at the carbon fiber cathode in the form of a very strong base with a pH of about 13. This strong base reacts with the polymer chain structure degrading the polymer. Polymers affected include bismaleimides (BMI), condensation polyimides, triazines, and blends thereof. Degradation occurs in the form of dissolved resin and loose fibers. The hydroxyl ions generated at the graphite cathode attack the O-C-N bond in the polyimide structure. Standard corrosion protection procedures were found to prevent polymer degradation under most conditions.

Chlorine-induced Cracking

Another highly reactive gas is chlorine, which will attack susceptible polymers such as acetal resin and polybutylene pipework. There have been many examples of such pipes and acetal fittings failing in properties in the US as a result of chlorine-induced cracking. In essence, the gas attacks sensitive parts of the chain molecules (especially secondary, tertiary, or allylic carbon atoms), oxidizing the chains and ultimately causing chain cleavage. The root cause is traces of chlorine in the water supply, added for its anti-bacterial action, attack occurring even at parts per million traces of the dissolved gas. The chlorine attacks weak parts of a product, and in the case of an acetal resin junction in a water supply system, it is the thread roots that were attacked first, causing a brittle crack to grow. Discoloration on the fracture surface was caused by deposition of carbonates from the hard water supply, so the joint had been in a critical state for many months. The problems in the US also occurred to polybutylene pipework, and led to the material being removed from that market, although it is still used elsewhere in the world.

chlorine attack of acetal resin plumbing joint

Biological Degradation

Biodegradable plastics can be biologically degraded by microorganisms to give lower molecular weight molecules. To degrade properly biodegradable polymers need to be treated like compost and not just left in a landfill site where degradation is very difficult due to the lack of oxygen and moisture.

Stabilizers

Hindered amine light stabilizers (HALS) stabilize against weathering by scavenging free radicals that are produced by photo-oxidation of the polymer matrix. UV-absorbers stabilizes against weathering by absorbing ultraviolet light and converting it into heat. Antioxidants stabilize the polymer by terminating the chain reaction due to the absorption of UV light from sunlight. The chain reaction initiated by photo-oxidation leads to cessation of crosslinking of the polymers and degradation the property of polymers.

UV Degradation

Many natural and synthetic polymers are attacked by ultraviolet radiation, and products using these materials may crack or disintegrate if they are not UV-stable. The problem is known as *UV degradation*, and is a common problem in products exposed to sunlight. Continuous exposure is a more serious problem than intermittent exposure, since attack is dependent on the extent and degree of exposure.

Many pigments and dyes can also be affected, and the problem known as phototendering can affect textiles such as curtains or drapes.

Susceptible Polymers

Effect of UV exposure on polypropylene rope

Common synthetic polymers that can be attacked include polypropylene and LDPE, where tertiary carbon bonds in their chain structures are the centres of attack. Ultraviolet rays interact with these bonds to form free radicals, which then react further with oxygen in the atmosphere, producing carbonyl groups in the main chain. The exposed surfaces of products may then discolour and crack, and in extreme cases, complete product disintegration can occur.

In fibre products like rope used in outdoor applications, product life will be low because the outer fibres will be attacked first, and will easily be damaged by abrasion for example. Discolouration of the rope may also occur, thus giving an early warning of the problem.

Polymers which possess UV-absorbing groups such as aromatic rings may also be sensitive to UV degradation. Aramid fibres like Kevlar, for example, are highly UV-sensitive and must be protected from the deleterious effects of sunlight.

Detection

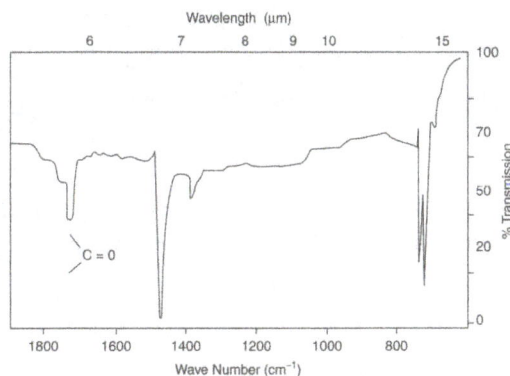

IR spectrum showing carbonyl absorption due to UV degradation of polyethylene

The problem can be detected before serious cracks are seen in a product using infrared spectroscopy, where attack occurs by oxidation of bonds activated by the UV radiation forming carbonyl groups in the polymer chains.

In the example shown at left, carbonyl groups were easily detected by IR spectroscopy from a cast thin film. The product was a road cone made by rotational moulding in LDPE, which had cracked prematurely in service. Many similar cones also failed because an anti-UV additive had not been used during processing. Other plastic products which failed included polypropylene mancabs used at roadworks which cracked after service of only a few months.

Prevention

UV attack by sunlight can be ameliorated or prevented by adding anti-UV chemicals to the polymer when mixing the ingredients, prior to shaping the product by injection moulding for example. UV stabilizers in plastics usually act by absorbing the UV radiation preferentially, and dissipating the energy as low-level heat. The chemicals used are similar to those in sunscreen products, which protect skin from UV attack. Frequently, glass can be a better alternative to polymers when it comes to UV degradation. Most of the commonly used glass types are highly resistant to UV radiation. Explosion protection lamps for oil rigs for example can be made either from polymer or glass. Here, the UV radiation and rough weathers belabor the polymer so much, that the material has to be replaced frequently.

Materials Testing

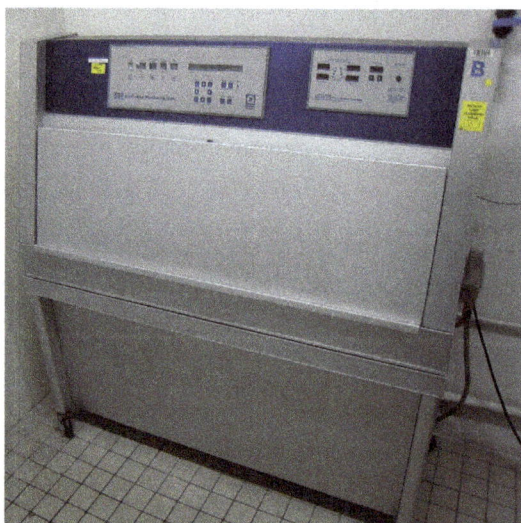
Example of test device

The effects of UV degradation on materials that require a long service life can be measured with accelerated exposure tests. With modern solar concentrator technologies, it is possible to simulate 63 years of natural UV radiation exposure on a test device in a single year.

Thermal Degradation of Polymers

Thermal degradation of polymers is molecular deterioration as a result of overheating. At high temperatures the components of the long chain backbone of the polymer can begin to separate (molecular scission) and react with one another to change the properties of the polymer. Thermal degradation can present an upper limit to the service temperature of plastics as much as the possibility of mechanical property loss. Indeed unless correctly prevented, significant thermal degradation can occur at temperatures much lower than those at which mechanical failure is likely to occur. The chemical reactions involved in thermal degradation lead to physical and optical property changes relative to the initially specified properties. Thermal degradation generally involves changes to the molecular weight (and molecular weight distribution) of the polymer and typical property changes include reduced ductility and embrittlement, chalking, color changes, cracking, general reduction in most other desirable physical properties. Thermal breakdown products may include a complex mixture of compounds, including but not limited to carbon monoxide, ammonia, aliphatic amines, ketones, nitriles, and hydrogen cyanide, which may be flammable, toxic and/or irritating. The specific materials generated will vary depending on the additives and colorants used, specific temperature, time of exposure and other immediate environmental factors.

The Mechanism of Thermal Degradation

Most types of degradation follow a similar basic pattern. The conventional model for thermal degradation is that of an auto-oxidation process which involves the major steps of initiation, propagation, branching, and termination.

Initiation

The initiation of thermal degradation involves the loss of a hydrogen atom from the polymer chain as a result of energy input from heat or light. This creates a highly reactive and unstable polymer 'free radical' (R•) and a hydrogen atom with an unpaired electron (H•).

Propagation

The propagation of thermal degradation can involve a variety of reactions and one of these is where the free radical (R•) reacts with an oxygen (O_2) molecule to form a peroxy radical (ROO•) which can then remove a hydrogen atom from another polymer chain to form a hydroperoxide (ROOH) and so regenerate the free radical (R•). The hydroperoxide can then split into two new free radicals, (RO•) + (•OH), which will continue to propagate the reaction to other polymer molecules. The process can therefore accelerate depending on how easy it is to remove the hydrogen from the polymer chain.

Termination

The termination of thermal degradation is achieved by 'mopping up' the free radicals to create inert products. This can occur naturally by combining free radicals or it can be assisted by using stabilizers in the plastic.

The Research Methods of Thermal Degradation of Polymers

TGA

(Thermogravimetric analysis) (TGA) refers to the techniques where a sample is heated in a controlled atmosphere at a defined heating rate whilst the samples mass is measured. When a polymer sample degrades, its mass decreases due to the production of gaseous products like carbon monoxide, water vapour and carbon dioxide.

DTA and DSC

(Differential thermal analysis) (DTA) and (differential scanning calorimetry) (DSC): Analyzing the heating effect of polymer during the physical changes in terms of glass transition, melting, and so on. These techniques measure the heat flow associated with oxidation.

Ways of Polymer Thermal Degradation

Depolymerisation

Under thermal effect, the end of polymer chain departs, and forms low free radical which has low activity. Then according to the chain reaction mechanism, the polymer loses the monomer one by one. However, the molecular chain doesn't change a lot in a short time. The reaction is shown below. This process is common for polymethymethacrylate (perspex).

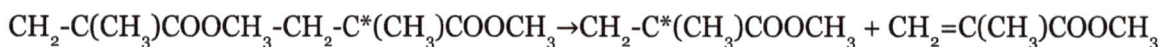

$$CH_2-C(CH_3)COOCH_3-CH_2-C^*(CH_3)COOCH_3 \rightarrow CH_2-C^*(CH_3)COOCH_3 + CH_2=C(CH_3)COOCH_3$$

Random Chain Scission

The backbone will break down randomly, this can occur at any position of the backbone, as a result the molecular weight decreases rapidly. As new free radicals with high reactivity are formed, monomers cannot be a product of this reaction, also intermolecular chain transfer and disproportion termination reactions can occur.

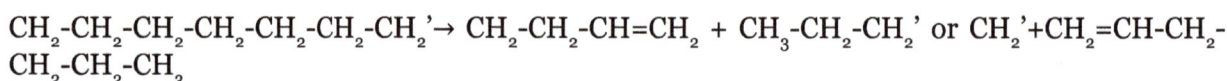

$$CH_2-CH_2-CH_2-CH_2-CH_2-CH_2-CH_2' \rightarrow CH_2-CH_2-CH=CH_2 + CH_3-CH_2-CH_2' \text{ or } CH_2'+CH_2=CH-CH_2-CH_2-CH_2-CH_3$$

Side-group Elimination

Groups that are attached to the side of the backbone are held by bonds which are weaker than the bonds connecting the chain. When the polymer is heated, the side groups are stripped off from the chain before it is broken into smaller pieces. For example the PVC eliminates HCl, under 100−120 °C.

Thermal Depolymerization

Thermal depolymerization (TDP) is a depolymerization process using hydrous pyrolysis for the reduction of complex organic materials (usually waste products of various sorts, often biomass and plastic) into light crude oil. It mimics the natural geological processes thought to be involved in the production of fossil fuels. Under pressure and heat, long chain polymers of hydrogen, oxygen, and carbon decompose into short-chain petroleum hydrocarbons with a maximum length of around 18 carbons.

Similar Processes

Thermal depolymerisation is similar to other processes which use superheated water as a major step to produce fuels, such as direct Hydrothermal Liquefaction. These are distinct from processes using dry materials to depolymerize, such as pyrolysis. The term Thermochemical Conversion (TCC) has also been used for conversion of biomass to oils, using superheated water, although it is more usually applied to fuel production via pyrolysis. Other commercial scale processes include the "SlurryCarb" process operated by EnerTech, which uses similar technology to decarboxylate wet solid biowaste, which can then be physically dewatered and used as a solid fuel called E-Fuel. The plant was designed to Rialto to process 683 tons of waste per day. However, it failed to perform to design standards and was closed down. The Rialto facility defaulted on its bond payments and is in the process of being liquidated. The Hydro Thermal Upgrading (HTU) process uses superheated water to produce oil from domestic waste. A demonstration plant is due to start up in The Netherlands said to be capable of processing 64 tons of biomass (dry basis) per day into oil. Thermal depolymerisation differs in that it contains a hydrous process followed by an anhydrous cracking / distillation process.

History

Thermal depolymerization is similar to the geological processes that produced the fossil fuels used today, except that the technological process occurs in a timeframe measured in hours. Until recently, the human-designed processes were not efficient enough to serve as a practical source of fuel—more energy was required than was produced.

The first industrial process to obtain gas, diesel fuels and other petroleum products through pyrolysis of coal, tar or biomass was designed and patented in the late 1920s by Fischer-Tropsch. In U. S. patent 2,177,557, issued in 1939, Bergstrom and Cederquist discuss a method for obtaining oil from wood in which the wood is heated under pressure in water with a significant amount of calcium hydroxide added to the mixture. In the early 1970s Herbert R. Appell and coworkers worked with hydrous pyrolysis methods, as exemplified by U. S. patent 3,733,255 (issued in 1973), which discusses the production of oil from sewer sludge and municipal refuse by heating the material in water, under pressure, and in the presence of carbon monoxide.

An approach that exceeded break-even was developed by Illinois microbiologist Paul Baskis in the 1980s and refined over the next 15 years (see U. S. patent 5,269,947, issued in 1993). The technology was finally developed for commercial use in 1996 by Changing World Technologies (CWT). Brian S. Appel (CEO of CWT) took the technology in 2001 and expanded and changed it into what is now referred to as TCP (Thermal Conversion Process), and has applied for and obtained several patents. A Thermal Depolymerization demonstration plant was completed in 1999 in Philadelphia by Thermal Depolymerization, LLC, and the first full-scale commercial plant was constructed in Carthage, Missouri, about 100 yards (91 m) from ConAgra Foods' massive Butterball turkey plant, where it is expected to process about 200 tons of turkey waste into 500 barrels (79 m^3) of oil per day.

Theory and Process

In the method used by CWT, the water improves the heating process and contributes hydrogen to the reactions.

In the Changing World Technologies (CWT) process, the feedstock material is first ground into small chunks, and mixed with water if it is especially dry. It is then fed into a pressure vessel reaction chamber where it is heated at constant volume to around 250 °C. Similar to a pressure cooker (except at much higher pressure), steam naturally raises the pressure to 600 psi (4 MPa) (near the point of saturated water). These conditions are held for approximately 15 minutes to fully heat the mixture, after which the pressure is rapidly released to boil off most of the water. The result is a mix of crude hydrocarbons and solid minerals. The minerals are removed, and the hydrocarbons are sent to a second-stage reactor where they are heated to 500 °C, further breaking down the longer hydrocarbon chains. The hydrocarbons are then sorted by fractional distillation, in a process similar to conventional oil refining.

The CWT company claims that 15 to 20% of feedstock energy is used to provide energy for the plant. The remaining energy is available in the converted product. Working with turkey offal as the feedstock, the process proved to have yield efficiencies of approximately 85%; in other words, the energy contained in the end products of the process is 85% of the energy contained in the inputs to the process (most notably the energy content of the feedstock, but also including electricity for pumps and natural gas or woodgas for heating). If one considers the energy content of the feedstock to be free (i.e., waste material from some other process), then 85 units of energy are made available for every 15 units of energy consumed in process heat and electricity. This means the "Energy Returned on Energy Invested" (EROEI) is (6.67), which is comparable to other energy harvesting processes. Higher efficiencies may be possible with drier and more carbon-rich feedstocks, such as waste plastic.

By comparison, the current processes used to produce ethanol and biodiesel from agricultural sources have EROEI in the 4.2 range, when the energy used to produce the feedstocks is accounted for (in this case, usually sugar cane, corn, soybeans and the like). These EROEI values are not directly comparable, because these EROEI calculations include the energy cost to produce the feedstock, whereas the above EROEI calculation for thermal depolymerization process (TDP) does not.

The process breaks down almost all materials that are fed into it. TDP even efficiently breaks down many types of hazardous materials, such as poisons and difficult-to-destroy biological agents such as prions.

Feedstocks and outputs with thermal depolymerization

Average TDP Feedstock Outputs				
Feedstock	**Oils**	**Gases**	**Solids (mostly carbon based)**	**Water (Steam)**
Plastic bottles	70%	16%	6%	8%
Medical waste	65%	10%	5%	20%
Tires	44%	10%	42%	4%
Turkey offal	39%	6%	5%	50%
Sewage sludge	26%	9%	8%	57%
Paper (cellulose)	8%	48%	24%	20%

(Note: Paper/cellulose contains at least 1% minerals, which was probably grouped under carbon solids.)

Carthage Plant Products

As reported on 04/02/2006 by Discover Magazine, a Carthage, Missouri plant was producing 500 barrels per day (79 m^3/d) of oil made from 270 tons of turkey entrails and 20 tons of hog lard. This represents an oil yield of 22.3 percent. The Carthage plant produces API 40+, a high value crude oil. It contains light and heavy naphthas, a kerosene, and a gas oil fraction, with essentially no heavy fuel oils, tars, asphaltenes or waxes. It can be further refined to produce No. 2 and No. 4 fuel oils.

TDP-40 Oil Classification by D-5443 PONA method	
Output Material	**% by Weight**
Paraffins	22%
Olefins	14%
Naphthenes	3%
Aromatics	6%
C14/C14+	55%
	100%

The fixed carbon solids produced by the TDP process have multiple uses as a filter, a fuel source and a fertilizer. It can be used as activated carbon in wastewater treatment, as a fertilizer, or as a fuel similar to coal.

Advantages

The process can break down organic poisons, due to breaking chemical bonds and destroying the molecular shape needed for the poison's activity. It is likely to be highly effective at killing pathogens, including prions. It can also safely remove heavy metals from the samples by converting them from their ionized or organometallic forms to their stable oxides which can be safely separated from the other products.

Along with similar processes, it is a method of recycling the energy content of organic materials without first removing the water. It can produce liquid fuel, which separates from the water physically without need for drying. Other methods to recover energy often require pre-drying (e.g. burning, pyrolysis) or produce gaseous products (e.g. anaerobic digestion).

Potential Sources of Waste Inputs

The United States Environmental Protection Agency estimates that in 2006 there were 251 million tons of municipal solid waste, or 4.6 pounds generated per day per person in the USA. Much of this mass is considered unsuitable for oil conversion.

Limitations

The process only breaks long molecular chains into shorter ones, so small molecules such as carbon dioxide or methane cannot be converted to oil through this process. However, the methane in the feedstock is recovered and burned to heat the water that is an essential part of the process. In addition, the gas can be burned in a combined heat and power plant, consisting of a gas turbine which drives a generator to create electricity, and a heat exchanger to heat the process input water from the exhaust gas. The electricity can be sold to the power grid, for example under a feed-in tariff scheme. This also increases the overall efficiency of the process (already said to be over 85% of feedstock energy content).

Another option is to sell the methane product as biogas. For example, biogas can be compressed, much like natural gas, and used to power motor vehicles.

Many agricultural and animal wastes could be processed, but many of these are already used as fertilizer, animal feed, and, in some cases, as feedstocks for paper mills or as boiler fuel. Energy crops constitute another potentially large feedstock for thermal depolymerization.

Current Status

Reports in 2004 claimed that the Carthage facility was selling products at 10% below the price of equivalent oil, but its production costs were low enough that it produced a profit. At the time it was paying for turkey waste.

The plant then consumed 270 tons of turkey offal (the full output of the turkey processing plant) and 20 tons of egg production waste daily. In February 2005, the Carthage plant was producing about 400 barrels per day (64 m³/d) of crude oil.

In April 2005 the plant was reported to be running at a loss. Further 2005 reports summarized some economic setbacks which the Carthage plant encountered since its planning stages. It was thought that concern over mad cow disease would prevent the use of turkey waste and other animal products as cattle feed, and thus this waste would be free. As it turned out, turkey waste may still be used as feed in the United States, so that the facility must purchase that feed stock at a cost of $30 to $40 per ton, adding $15 to $20 per barrel to the cost of the oil. Final cost, as of January 2005, was $80/barrel ($1.90/gal).

The above cost of production also excludes the operating cost of the thermal oxidizer and scrubber added in May 2005 in response to odor complaints.

A biofuel tax credit of roughly $1 per US gallon (26 ¢/L) on production costs was not available because the oil produced did not meet the definition of "biodiesel" according to the relevant American tax legislation. The Energy Policy Act of 2005 specifically added thermal depolymerization to a $1 renewable diesel credit, which became effective at the end of 2005, allowing a profit of $4/barrel of output oil.

Company Expansion

The company has explored expansion in California, Pennsylvania, and Virginia, and is presently examining projects in Europe, where animal products cannot be used as cattle feed. TDP is also being considered as an alternative means for sewage treatment in the United States.

Smell Complaints

The pilot plant in Carthage was temporarily shut down due to smell complaints. It was soon restarted when it was discovered that few of the odors were generated by the plant. Furthermore, the plant agreed to install an enhanced thermal oxidizer and to upgrade its air scrubber system under a court order. Since the plant is located only four blocks from the tourist-attracting town center, this has strained relations with the mayor and citizens of Carthage.

According to a company spokeswoman, the plant has received complaints even on days when it is not operating. She also contended that the odors may not have been produced by their facility, which is located near several other agricultural processing plants.

On December 29, 2005, the plant was ordered by the state governor to shut down once again over allegations of foul odors as reported by MSNBC.

As of March 7, 2006, the plant has begun limited test runs to validate it has resolved the odor issue.

As of August 24, 2006, the last lawsuit connected with the odor issue has been dismissed and the problem is acknowledged as fixed. In late November, however, another complaint was filed over bad smells. This complaint was closed on January 11 of 2007 with no fines assessed.

Status as of February 2009

A May 2003 article in Discover magazine stated, "Appel has lined up federal grant money to help build demonstration plants to process chicken offal and manure in Alabama and crop residuals and grease in Nevada. Also in the works are plants to process turkey waste and manure in Colorado and pork and cheese waste in Italy. He says the first generation of depolymerization centers will be up and running in 2005. By then it should be clear whether the technology is as miraculous as its backers claim."

However, as of August 2008, the only operational plant listed at the company's website is the initial one in Carthage, Missouri.

Changing World Technology applied for an IPO on August 12; 2008, hoping to raise $100 million.

The unusual Dutch Auction type IPO failed possibly because CWT has lost nearly $20 million with very little revenue.

CWT, the parent company of Renewable Energy Solutions, filed for Chapter 11 bankruptcy. No details on plans for the Carthage plant have been released.

In April 2013, CWT was acquired by a Canadian firm, Ridgeline Energy Services, based in Calgary.

Similar Technologies

- Plasma Converters use powerful electric arcs to reduce and extract energy from waste.
- Wet oxidation
- Hydrocracking

Chemically Assisted Degradation of Polymers

Chemically assisted degradation of polymers is a type of polymer degradation that involves a change of the polymer properties due to a chemical reaction with the polymer's surroundings. There are many different types of possible chemical reactions causing degradation however most of these reactions result in the breaking of double bonds within the polymer structure.

Examples of Chemically Assisted Degradation

Degradation of Rubber by Ozone

One common example of chemically assisted degradation is the degradation of rubber by ozone particles. Ozone is a naturally occurring atmospheric molecule that is produced by electric discharge or through a reaction of Oxygen with solar radiation. Ozone is also produced with atmospheric pollutants reacted with ultraviolet radiation. For a reaction to occur, ozone concentrations only have to be as low as 3-5 parts per hundred million (pphm) and when these concentrations are reached, a reaction occurs with a thin surface layer (5 x10-7 metres) of the material. The ozone molecules react with the rubber which in most cases is unsaturated (contains double bonds), however a reaction will still occur in saturated polymers (those containing only single bonds). When reaction occurs, scission of the polymer chain (breaking of double covalent bonds) takes place forming decomposition products:

Chain scission increases with the presence of active Hydrogen molecules (for example, in water) as well as acids and alcohols. Along with this type of reaction, cross linking and side branch formations also occur by an activation of the double bond and these make the rubber material more brittle. Due to the increase in brittleness due to the chemical reactions, cracks form in areas of high stress. As propagation of these cracks increases, new surfaces are opened for degradation to occur.

Degradation of Poly(Vinyl) Chloride (PVC)

Degradation can also occur as a result of the formation, and then breakage of double bonds, such as solvolysis in PVC(Peacock). Solvolysis occurs when a Carbon-X bond, with X representing a halogen, is broken. This occurs in PVC in the presence of an acid species. Active Hydrogen atoms will remove a Chlorine atom from the polymer molecule, forming Hydrochloric acid (HCl). The HCl produced may then cause dechlorination of adjacent Carbon atoms. The dechlorinated Carbon atoms then tend to form double bonds, which can be attacked and broken by ozone, just like the degradation of rubbers described above.

Degradation of Polyester

The degradation of a polyester may occur without the presence of the acidic catalyst that causes degradation of PVC. During hydrolysis water acts as the reactive catalyst instead of the acid. It causes degradation mainly at high temperature and pressure during processing.

In this process the water molecule will attack the C-O ester bond, splitting the polymer in half. The water molecule will then dissociate, with one Hydrogen atom forming a carboxylic acid group on the Carbon atom with the double bonded Oxygen, while the remaining atoms form an alcohol on the other chain end. These reactive products may also cause further degradation of the polymer chain. This chain scission lowers average molecular weight of the polymer, decreasing the number and strength of intermolecular bonds as well as the degree of entanglement. This will increase chain mobility, decreasing strength of the polymer and increasing deformation at low stresses.

Protection Against Chemically Assisted Degradation

Both physical and chemical barriers can be used to protect a polymer from chemically assisted degradation. A physical barrier must provide continuous protection, must not react with the polymer's environment, must be flexible so that stretching may occur and must also be able to regenerate (after wear processes). A chemical barrier must be highly reactive with the polymer's surroundings so that the barrier reacts with the environmental conditions rather than the polymer itself. This barrier involves addition of a material into the polymer blend during fabrication of the polymer. Due to this, the barrier addition must have a suitable solubility, must be economically feasible and must not hinder the production process. For the barrier to be activated, the addition must diffuse to the surface and so a suitable diffusivity is also required. There are four theories on how these types of barriers protect the polymer material:

- *Scavenger theory*: the protective layer reacts with the ozone rather than the polymer.

- *Protective film theory*: the protective layer reacts with the polymer producing a thin film on the polymer surface which is inert and can't be penetrated.

- *Re-linking theory*: the protective layer causes broken double bonds to be reformed.

- *Self-healing theory*: the protective layer reacts with degraded polymer chains to form low-molecular-weight material which forms an inert film on the surface.

Of these theories, the scavenger theory is the most common and most important. However, more than one theory can act at the same time and the theory that takes place depends on the protective materials, the polymer and surrounding environment.

Photo-oxidation of Polymers

Photo-oxidation is the degradation of a polymer surface in the presence of oxygen or ozone. The effect is facilitated by radiant energy such as UV or artificial light. This process is the most significant factor in weathering of polymers. Photo-oxidation is a chemical change that reduces the polymer's molecular weight. As a consequence of this change the material becomes more brittle, with a reduction in its tensile, impact and elongation strength. Discoloration and loss of surface smoothness accompany photo-oxidation. High temperature and localized stress concentrations are factors that significantly increase the effect of photo-oxidation.

Photo-oxidation Protection

Poly(ethylene-naphthalate) (PEN) can be protected by applying a zinc oxide coating, which acts as protective film reducing the diffusion of oxygen. Zinc oxide can also be used on polycarbonate (PC) to decrease the oxidation and photo-yellowing rate caused by solar radiation.

Effects of Dyes/Pigments

Adding pigment light absorbers and photostabilizers (UV absorbers) is one way to minimise photo-oxidation in polymers. Antioxidants are used to inhibit the formation of hydroperoxides in the photo-oxidation process.

Dyes and pigments are used in polymer materials to provide color changing properties. These additives can reduce the rate of polymer degradation. Cu-phthalocyanine dye can help stabilize against degradation, but in other situations such as photochemical aging can actually accelerate degradation. The excited Cu-phthalocyanine may abstract hydrogen atoms from methyl groups in the PC, which increase the formation of free radicals. This acts as the starting points for the sequential photo-oxidation reactions leading to the degradation of the PC.

Electron transfer sensitization is a mechanism where the excited Cu-phthalocyanine abstracts electrons from PC to form Cu-Ph radical anion and PC radical cations. These species in the presence of oxygen can cause oxidation of the aromatic ring.

Chemical Mechanism

Aldehydes, ketones and carboxylic acids along or at the end of polymer chains are generated by oxygenated species in photolysis of photo-oxidation. The initiation of photo-oxidation reactions is due to the existence of chromophoric groups in the macromolecules. Photo-oxidation can occur simultaneously with thermal degradation and each of these effects can accelerate the other.

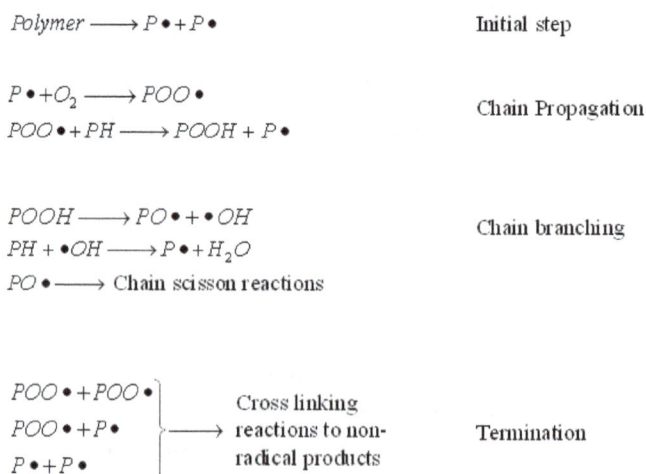

$$Polymer \longrightarrow P\bullet + P\bullet \qquad\qquad \text{Initial step}$$

$$P\bullet + O_2 \longrightarrow POO\bullet$$
$$POO\bullet + PH \longrightarrow POOH + P\bullet \qquad \text{Chain Propagation}$$

$$POOH \longrightarrow PO\bullet + \bullet OH$$
$$PH + \bullet OH \longrightarrow P\bullet + H_2O \qquad\qquad \text{Chain branching}$$
$$PO\bullet \longrightarrow \text{Chain scisson reactions}$$

$$\left. \begin{array}{l} POO\bullet + POO\bullet \\ POO\bullet + P\bullet \\ P\bullet + P\bullet \end{array} \right\} \longrightarrow \begin{array}{l} \text{Cross linking} \\ \text{reactions to non-} \\ \text{radical products} \end{array} \qquad \text{Termination}$$

where PH = Polymer

P• = Polymer alkyl radical

PO• = Polymer oxy radical (Polymer alkoxy radical)

POO• = Polymer peroxy radical (Polymer alkylperoxy radical)

POOH = Polymer hydroperoxide

HO• = hydroxy radical

The photo-oxidation reactions include chain scission, cross linking and secondary oxidative reactions. The following process steps can be considered:

1. Initial step: Free radicals are formed by photon absorption.

2. Chain Propagation step: A free radical reacts with oxygen to produce a polymer peroxy radical (POO•). This reacts with a polymer molecule to generate polymer hydroperoxide (POOH) and a new polymer alkyl radical (P•).

3. Chain Branching: Polymer oxy radicals (PO•) and hydroxy radicals (HO•) are formed by photolysis.

4. Termination step: Cross linking is a result of the reaction of different free radicals with each other.

References

- Faudree, Michael C. (1991). "Relationship of Graphite/Polyimide Composites to Galvanic Processes" (PDF). Society for the Advancement of Material and Process Engineering (SAMPE) Journal. 2: 1288–1301. ISBN 0-938994-56-5.

- "Photo-oxidisation of electroluminescent polymers studied by core level photoabsorption specttroscopy" (PDF). American institute of physics 1996. Retrieved 9 February 2011.

- "THE PHOTO-OXIDATION OF POLYMERS - A comparison with low molecular weight compounds" (PDF). Pergamon Press Ltd. 1979 - Pure & Appi. Chem., Vol. 51, pp.233–240. Retrieved 9 February 2011.

Permissions

We would like to thank the editorial team for lending their expertise to make the book truly unique. They have played a crucial role in the development of this book. Without their invaluable contributions this book wouldn't have been possible. They have made vital efforts to compile up to date information on the varied aspects of this subject to make this book a valuable addition to the collection of many professionals and students.

This book was conceptualized with the vision of imparting up-to-date and integrated information in this field. To ensure the same, a matchless editorial board was set up. Every individual on the board went through rigorous rounds of assessment to prove their worth. After which they invested a large part of their time researching and compiling the most relevant data for our readers.

The editorial board has been involved in producing this book since its inception. They have spent rigorous hours researching and exploring the diverse topics which have resulted in the successful publishing of this book. They have passed on their knowledge of decades through this book. To expedite this challenging task, the publisher supported the team at every step. A small team of assistant editors was also appointed to further simplify the editing procedure and attain best results for the readers.

Apart from the editorial board, the designing team has also invested a significant amount of their time in understanding the subject and creating the most relevant covers. They scrutinized every image to scout for the most suitable representation of the subject and create an appropriate cover for the book.

The publishing team has been an ardent support to the editorial, designing and production team. Their endless efforts to recruit the best for this project, has resulted in the accomplishment of this book. They are a veteran in the field of academics and their pool of knowledge is as vast as their experience in printing. Their expertise and guidance has proved useful at every step. Their uncompromising quality standards have made this book an exceptional effort. Their encouragement from time to time has been an inspiration for everyone.

The publisher and the editorial board hope that this book will prove to be a valuable piece of knowledge for students, practitioners and scholars across the globe.

Index

www.ingramcontent.com/pod-product-compliance
Lightning Source LLC
Chambersburg PA
CBHW061255190326
41458CB00011B/3677

* 9 7 8 1 6 3 5 4 9 2 2 8 6 *